Preventing Firefighter Disorientation

PREVENTING FIREFIGHTER
DISORIENTATION
Enclosed Structure Tactics for the Fire Service

WILLIAM R. MORA

> **Disclaimer**
> The recommendations, advice, descriptions, and the methods in this book are presented solely for educational purposes. The author and publisher assume no liability whatsoever for any loss or damage that results from the use of any of the material in this book. Use of the material in this book is solely at the risk of the user.

Copyright© 2016 by
PennWell Corporation
1421 South Sheridan Road
Tulsa, Oklahoma 74112-6600 USA

800.752.9764
+1.918.831.9421
sales@pennwell.com
www.FireEngineeringBooks.com
www.pennwellbooks.com
www.pennwell.com

Marketing Manager: Sarah De Vos
National Account Manager: Cindy J. Huse

Director: Mary McGee
Managing Editor: Marla Patterson
Production Manager: Sheila Brock
Production Editor: Tony Quinn
Cover Designer: Charles Thomas

Library of Congress Cataloging-in-Publication Data

Names: Mora, William R., 1953- author.
Title: Preventing firefighter disorientation : enclosed structure tactics for the fire service / William R. Mora.
Description: Tulsa, Oklahoma : PennWell Corporation, [2016] | Includes bibliographical references and index.
Identifiers: LCCN 2016000793 | ISBN 9781593703745
Subjects: LCSH: Fire extinction. | Command and control at fires. | Buildings--Evacuation.
Classification: LCC TH9145 .M67 2016 | DDC 628.9/25--dc23
LC record available at http://lccn.loc.gov/2016000793

All rights reserved. No part of this book may be reproduced, stored in a retrieval system, or transcribed in any form or by any means, electronic or mechanical, including photocopying and recording, without the prior written permission of the publisher.

Printed in the United States of America

1 2 3 4 5   20 19 18 17 16

In honor of all fallen firefighters

Contents

Preface xiii
Acknowledgments xv

1 The Structural Firefighter Fatality Problem 1
    The U.S. Firefighter Disorientation Study 1979–2001 6
        Methodology 6
    Firefighter Disorientation 8
    References 10

2 Terms, Risk, and Assumptions 11
    Opened Structure Defined 12
    Enclosed Structure Defined 21
    Degrees of Danger 25
        Opened structure assumptions 26
        The worst-case scenario assumption for opened structures 26
        Enclosed structure assumptions 27
        The worst-case scenario assumption for enclosed structures 27
        The large structure assumption 28
        The tall structure assumption 30
        The fall height criteria 32
        The unprotected and protected enclosed structure
            assumptions 35
    The 1991 One Meridian Plaza Incident 36
    The 1988 First Interstate Tower Incident 37
    The 1987 Fall River Incident 39
    Using the Terms "Opened" or "Enclosed" for Added
        Safety 43
    References 46

3 Types of Enclosed Structures 47
    Enclosed Structure Characteristics 47
    Opened Structure with a Basement 50
        Common signs of a basement 54
    Enclosed Structure with a Basement 59
    Reduced Situational Awareness During Basement Fires 62

Opened and Enclosed Structure Combination 64
Enclosed Spaces 65
    High-rise hallways 66
    High-rise stairwells. 66
Large Enclosed Structure Defined 67
Large Enclosed Structure Risk Classification Criteria. 69
References 76

4 Types of Obscured Visibility Conditions and Types of Firefighter Disorientation 79
Zero Visibility Conditions 79
Prolonged Zero Visibility Conditions (PZVCs) 81
Types of Firefighter Disorientation 82
    Disorientation secondary to prolonged zero visibility
        conditions (PZVCs) 82
    Disorientation secondary to flashover. 84
    Disorientation secondary to backdraft 87
    Disorientation secondary to collapse 88
    Disorientation secondary to wind-driven fire 89
    Disorientation secondary to conversion steam 91
References 92

5 Similarities in Disorientation Fires 93
References 98

6 The Firefighter Disorientation Sequence 99
The Out-of-Sequence Chain. 100
    Firefighter disorientation may occur in typical sequence 101
    Firefighter disorientation may occur out of sequence 102
References 104

7 Initial Size-Up Factors. 105
Misreading Initial Size-Up Factors. 105
Analysis of the Structural Firefighter Fatality Database. 108
    Background 109
    Information sought. 109
    Criteria. 109
    Color coding for structure type, action, and outcome. 110

Findings 110
Conclusion 111
References 111

8 Avoiding Life-Threatening Hazards 113
Backdraft 113
Actions to avoid backdraft 115
Exterior and interior signs of backdraft 116
The Colorado explosion incident 117
The Crooksville, Ohio, backdraft incident 118
The Memphis backdraft incident 120
The Boston backdraft incident 121
The Chicago backdraft incident 123
Flashover 124
Avoiding the flashover hazard 126
Flashover contingency action plans 128
Balanced concern for water damage 129
Fire dynamics 131
Flow path 132
Ventilation-induced flashover 133
Softening the target 134
Self-Venting and Wind-Driven Fires 139
Wind control devices (WCDs) 140
Avoiding wind-driven fires 142
Size-up factors for high-rise, wind-driven fires 143
Door control 144
Doors, walls, and ceilings serving as wind control devices
 and thermal barriers 146
High-rise wind-driven fire tactics 147
Obscured vision and signs of breathing difficulty 153
The Oak Park, Michigan, incident 154
The angle of the smoke plume 156
The wind-driven fire action plan 159
Rapid fire spread from above 164
Rapid fire spread from below 167
Disorienting hazards associated with high-rise fires 168

> Rollover 170
> Floor Collapse 172
> Prolonged Zero-Visibility Conditions 173
> Preplanning Unprotected Enclosed Structures 175
> References 177

9 Hose Evolutions and Fireground Realities 181
> Traditional Approaches to Fire Flow Development 184
>> The attack-supply evolution (blitz attack) 184
>> The single-engine evolution 186
> Contemporary Approaches to Fire Flow Development 188
>> The dual pumping evolution 188
>> Dual pumping at a hydrant with 2½-in. or 3-in. supply hose 189
>> Dual pumping at the scene with 5-in. hose 190
>> The single-engine evolution with dual pumping at the scene . . . 195
>> The attack-supply evolution with dual pumping at the scene (supplying with straight and reverse lays) 197
>> Dual pumping at the scene during the attack-supply evolution with a three-engine response 199
>> The double attack-supply evolution: immediately implementing dual pumping at the scene with a three-engine response 200
>> Fire flow capability of the double attack-supply evolution 204
> Safe Operating Residual Pressure 208
> Benefits of the Double Attack-Supply Evolution 211
>> The quick backup line 211
>> Portable interior monitors 214
> Greatest Levels of Effectiveness 216
> The Hydrant Flowchart for Fire Flow Determination 217
> References 220

10 Hazards of Construction 221
> References 228

11 Coordinated Ventilation 229
> Types of Ventilation 229
>> Horizontal ventilation 229
>> Vertical ventilation 229

	Hydraulic ventilation 230
	Principles of Ventilation 230
	Coordinating Ventilation............................. 231
	References 233

12 Enclosed Structure Tactics and Guidelines 235

Tactics for Opened Structures with a Basement 236
    Overview 236
    Tactics for unoccupied opened structure basement fires 244
    Tactics for occupied opened structure basement fires 248
    Capability of piercing nozzles 255

Large Enclosed Structure Tactics: Common Operational and
    Safety Aspects of the Unknown–Known Guidelines ... 256
    The initial large enclosed structure size-up................. 257
    The cautious interior assessment 258
    The interior attack from the original point of entry 268
    The short interior attack 268
    The defensive attack 274

Large Enclosed Structure Tactics: Using the U–K Method
    for Decision Making (Unknown–Known Location of
    the Fire Scenarios) 275
    Condition 1: smoke is dispersed and location of the seat
        of the fire is unknown 277
    Condition 2: smoke is concentrated in one area and the
        seat of the fire is known 286
    Quickly maximizing the flow 290

Runoff in Residential versus Large Enclosed Commercial
    Structure Fires.................................... 292

Primary Search Requirements in Residential versus
    Nonresidential Large Enclosed Structures 293

References 294

13 Firefighter Disorientation Case Reviews:
Considering Enclosed Structure Tactics 295

The Worcester Incident............................... 296
    The Worcester incident analysis 298

Preventing Firefighter Disorientation

  The Worcester incident with enclosed structure tactics 298
 The Memphis Incident 301
  The Memphis incident analysis 302
  The Memphis incident with enclosed structure tactics 303
 The Carthage Incident 305
  The Carthage incident analysis 307
  The Carthage incident with enclosed structure tactics 307
 The Phoenix Incident 310
  The Phoenix incident analysis 314
  The Phoenix incident with enclosed structure tactics 315
  Option 1 316
  Option 2 317
  Option 3 318
 The Charleston Incident 320
  The Charleston incident analysis 326
  The Charleston incident with enclosed structure tactics 328
  Option 1 330
  Option 2 331
  Option 3 332
  Option 4 333
 Dual Pumping at a Distant Hydrant with 5-in. Hose 340
 Effectiveness of Vertical Ventilation 343
 "Fire through the Roof" Tactics 345
 References 347

14 Summary and Conclusion 349

 Appendix 355
  Unprotected Enclosed Structures: A Global Problem 355
  Reference 357

 Glossary 359

 Index 363

Preface

The "U.S. Firefighter Disorientation Study 1979–2001" examined 17 national structure fires that resulted in 23 line-of-duty deaths. The study determined that in 100% of the cases analyzed, firefighters experienced disorientation in structures and spaces that specifically had an enclosed architectural design. A follow-up study of 444 structural firefighter fatalities occurring over a 16-year time span (1990–2006) further revealed the degree of danger associated with enclosed structure fires and with the strategy and tactics used at the time. The results showed that traumatic fatalities occurred at the scene of enclosed structure fires at a disproportionate rate when traditional fast and aggressive interior attacks were used. A more immediate concern is that deaths and injuries attributable to firefighter disorientation and/or associated with enclosed structure fires continue to take place by firefighters relying on inaccurate size-up factors and on traditional strategy and tactics routinely implemented by the vast majority of fire departments in the country. Since studies have shown it is very difficult to rescue disoriented or trapped firefighters from deep interior locations in a timely manner, prevention is the key. Therefore, if prevention of injuries and deaths in which disorientation is a contributing factor is to be achieved nationally, sweeping change in the tactics currently used at enclosed structure fires must occur. The information to help reach that goal is offered here in the form of terms, tactics, and assumptions and through the use of available technology that collectively produce a different approach to safely manage extremely dangerous enclosed structure fires.

Acknowledgments

It is impossible for any one person to write a book that focuses on the prevention of firefighter disorientation and fatalities. It actually takes not only a considerable amount of time and effort but also assistance and guidance of others. For their support, thanks go to the individuals, businesses, and organizations who willingly offered their time, information, and insight to help make this book understandable and hopefully effective. I would like to begin by offering my thanks to the investigators of the National Institute for Occupational Safety and Health (NIOSH) for the high-quality data they have collected and distributed through the years. This book relies on NIOSH publications extensively. Additionally, I thank Dan Madrzykowski of the National Institute of Standards and Technology (NIST) for his dedication, assistance, and considerable contribution by sharing years of research-based tactical and technical information and doing so during his personal time. I also wish to thank Steve Kerber with Underwriters Laboratory for sharing tactical information based on experimentation and science in terms that can easily be converted to action on the fireground. I also would like to thank Tim Merinar of NIOSH for his patience and considerable help in providing specific case photos suitable for firefighter review. Quality photography is essential when attempting to explain fireground operations, and so my thanks go to the following individuals and organizations for their assistance: Lindsay Ackermann, Chief Gary Bowker, Lt. Clifford Clement, Jason Denny, Stewart English, Dan Folk, Scott LaPrade, Peter Matthews, Captain Dennis Meier, Lt. Roger Mora, Tim Olk, Professor Mike Pickett, Paul Ramirez, Crew Commander Eddie Robertson, Andy Thomas, and Firefighter Dennis Walus.

Additional thanks go to members of the following organizations: the U.S. Fire Administration (USFA); National Fire Data Center, including Mark Whitney, Bill Troup, and Kenneth Kuntz; National Fire Protection Association (NFPA); National Fallen Firefighters Foundation (NFFF); International Fire Service Training Association (IFSTA); International Association of Fire Chiefs—Safety, Health and Survival Section (IAFC-SHS); International Fire Service Journal of Leadership and Management (IFSJLM); Atlanta Fire Department; Wharton (Texas)

Fire Department; Carthage (Missouri) Fire Department; Memphis Fire Department; Worcester (Massachusetts) Fire Department; Charleston Fire Department; Houston Fire Department; Phoenix Fire Department; Salisbury Fire Department; Pratt (Kansas) Fire Department; Wichita (Kansas) Fire Department; New Jersey Division of Fire Safety; U.S. Bureau of Alcohol, Tobacco, Firearms and Explosives (ATF); Texas State Fire Marshal's Office; and West Virginia State Fire Marshal's Office. Special thanks go to participants of the NIST–UL fire studies, including the International Society of Fire Service Instructors (ISFSI) and the Toledo, Chicago, Spartanburg, and New York City Fire Departments. I am appreciative of aerial photography shared by Pictometry International Corporation and offer acknowledgments to Tony Quinn, production editor, and Marla Patterson, managing editor, of PennWell Publishing for guidance during the process. For those overlooked, I extend my sincerest appreciation.

I would also like to thank former and current administrations and members of the San Antonio Fire Department for always being willing to try new methods of operation for enhancing safety and effectiveness on the fireground. In addition, gratitude is given to former fire engineer Robert Jacinto of the San Antonio Fire Department for sharing his experience with disorientation. To conclude, I wish to thank my wife, Linda, and son, Greg, for their unwavering support and encouragement during the course of this lengthy project. One final and very special thanks goes to my daughter, Nicole Mora Wong, who was always there to share her technical talents, knowledge of effective communication, and sound advice whenever I needed it the most.

1

THE STRUCTURAL FIREFIGHTER FATALITY PROBLEM

Although firefighters manage the vast majority of structure fires safely and effectively, the stark reality is that firefighter fatalities at the scene of structure fires continue to be a constant occupational risk in the American fire service. The heavy losses suffered in terms of fatalities and injuries by all firefighters nationwide have been devastating for survivors, departments, and the communities they served.

Factors involving one specific injury incident in 1997 served as the impetus to more closely examine the risk of disorientation. During a structure fire at a thrift store, an acting lieutenant of the San Antonio Fire Department became disoriented after he was separated from his handline and crew. Low on air, in blinding smoke, and with fire rolling overhead, the officer tried to evacuate toward what he thought was the point of entry but mistakenly crawled in the opposite direction, deeper into the large commercial structure. As the structure flashed over, he dove through a window that had been forced open only minutes earlier by a truck crew. Though suffering critical thermal injuries, he survived. Additionally, nine other disoriented firefighters were barely able to evacuate before running out of air and before flashover.

While every firefighter fatality in every category is considered a catastrophic event, the traumatic fatalities that take place during the course of structural firefighting are the focus here. Additionally, through the combined, ongoing efforts of several safety-oriented organizations, additional losses may have already been prevented, as indicated by the most recent fatality reports. The following are some of the organizations that are involved with firefighter safety and survival:

- National Fire Protection Association (NFPA)
- U.S. Fire Administration (USFA)
- International Association of Fire Chiefs (IAFC) Safety Health and Survival Section
- International Association of Fire Fighters (IAFF)
- National Fallen Firefighters Foundation (NFFF) Everyone Goes Home Firefighter Life Safety Initiatives (FLSI) Program
- National Institute of Standards and Technology (NIST)
- National Institute for Occupational Safety and Health (NIOSH) Fire Fighter Fatality Investigation and Prevention Program (FFFIPP)

It is difficult to determine which firefighter fatality prevention effort has played the greatest part in helping to reduce the number of on-duty firefighter deaths in recent years or if it has been a collective effort. However, it is encouraging to note the number of total firefighter deaths in 2014 was fortunately down once again. The report titled "Firefighter Fatalities in the United States—2014," by Rita F. Fahy, Paul R. LeBlanc, and Joseph L. Molis of the NFPA's Fire Analysis and Research Division, states that "In 2014, 64 firefighters died while on-duty in the U.S. This total represents a significant decrease from the 97 deaths that occurred in 2013, when three incidents alone claimed a total of 32 lives." Concerning the distribution of the 64 deaths by type of duty, "the largest share of deaths occurred while firefighters were operating at fires (22 deaths), accounting for about one-third of the on-duty deaths in 2014." The report goes on to state that although fatalities occurred at various types of fireground incidents including wildland and vehicle fires, "Seventeen of the 22 fire ground deaths occurred at 15 structure fires. Five firefighters were caught or trapped by rapid fire progress (including flashover) in three incidents. Structural collapses resulted in two deaths. In separate incidents, three firefighters became lost inside structures and three died in fatal falls." Finally, and according to the

NFPA report, "None of the structures in which firefighters died was reported to have had an automatic fire suppression system." In light of these findings, it is obvious that firefighters today should be especially concerned about their safety, specifically at the scene of structure fires.

The following case summaries focusing exclusively on traumatic firefighter fatalities at structure fires serve as clear evidence of the ongoing structural firefighter fatality problem.

- On July 3, 2010, two Texas volunteer fire captains became disoriented after making a fast and aggressive interior attack into a large commercial structure. One officer died, and the second was able to barely escape the heavily involved structure.
- In December 2010, two Chicago firefighters lost their lives during the late stages of a large abandoned commercial structure fire when the roof they were standing on collapsed. The fire also injured over a dozen firefighters.
- On January 19, 2011, a Maryland volunteer firefighter died while conducting a primary search after being caught in a rapid fire progression at a three-story apartment building.
- On June 15, 2011, an Illinois volunteer fire officer with the Muncie Fire Department died and two others were injured while making an aggressive interior attack during the course of a wind-driven, unoccupied, large structure fire involving a place of worship. A roof collapse trapped the firefighter in the heavily involved structure.
- On July 28, 2011, a career captain with the Asheville (North Carolina) Fire Department was conducting a primary search in a five-story office building when he became disoriented in heavy smoke. Although rapid intervention team members were able to locate and remove the captain from the structure, he unfortunately died of his injuries. Other firefighters involved in the operation also sustained injuries.
- On March 4, 2012, a Wisconsin volunteer lieutenant lost his life at a theater fire after a bowstring truss roof collapsed, trapping him within the structure.

- On February 15, 2013, first-arriving career firefighters with the Bryan (Texas) Fire Department executed a fast attack into a large unoccupied structure involving a place of assembly. One officer became disoriented and notified others he was running low on air. During the rescue attempt, another lieutenant working as part of a rapid intervention team was exposed to heavy fire conditions. Both officers succumbed to their injuries, and two additional firefighters suffered burn injuries.
- On May 8, 2013, a Wayne Westland (Illinois) firefighter died after a roof collapse during a strip mall fire involving a billiard hall.
- On May 31, 2013, four firefighters of the Houston Fire Department were conducting a primary search when they suddenly became trapped by a collapsing roof and died. Thirteen other firefighters were also injured during the five-alarm fire, which destroyed the heavily involved hotel.
- On January 26, 2014, during a Mayday incident, a rapidly spreading fire in a small two-story commercial-residential occupancy claimed the lives of two Toledo firefighters as they conducted a primary search.

In the U.S. fire service today, traumatic and preventable line-of-duty deaths on the scene of unprotected enclosed structure fires continue to unfold, and much more must and can be done to prevent them.

As the effort to reach higher levels of fireground safety advances, additional improvements in training and in personal protective and safety equipment, as well as a major shift in operating procedures, are needed in order to minimize traumatic structural firefighter fatalities. To that end, the ultimate challenge facing today's firefighters is learning from and transforming the difficult lessons of the past into safer management of enclosed structure fires for the future (fig. 1–1).

Fig. 1–1. In July 2010, two volunteer captains of the Wharton, Texas, Fire Department became disoriented in zero-visibility conditions at the scene of an unprotected, enclosed, large structure fire. The fire took place at the Maxim Egg Farm located in Bolin, Texas. One disoriented officer, who had become separated from another officer and the handline, exhausted his air supply and barely escaped with his life during the incident. After getting the attention of exterior firefighters by banging on the structure's interior metal siding, firefighters cut a hole and opened the structure's wall from the exterior to allow for his escape. The second disoriented officer, however, was not found in time, as deteriorating conditions prevented entry by rapid intervention teams. (Courtesy of the U.S. Bureau of Alcohol, Tobacco, Firearms and Explosives [ATF] and Texas State Fire Marshal's Office Investigation Case FY 10-01, http://www.tdi.texas.gov/reports/fire/documents/fmloddaraguz.pdf.)

The U.S. Firefighter Disorientation Study 1979–2001

The "U.S. Firefighter Disorientation Study 1979–2001" (Mora 2003) was a three-year study that examined 17 structure fires occurring over a 22-year time span. The specific fires, taking place in communities located across the country, were selected for analysis since they all had one known factor in common: Firefighters became disoriented during the course of each incident. Tragically, these specific fires resulted in 23 firefighters losing their lives as well as numerous injuries and several narrow escapes during hazardous interior firefighting operations.

Methodology

The disorientation study examined data that included safety reports, tactical work sheets, videos, and interviews with firefighters actively involved in the incidents. An investigative report from the Texas State Fire Marshal's Office and several from the FFFIPP conducted by investigators of NIOSH were also closely studied and proved to be helpful during the examination of different aspects of each incident. More specifically, the reports by NIOSH and the Texas Fire Marshal provided comprehensive facts concerning each department, structure, staffing, training, equipment, the weather, and a timeline of actions and conditions as they unfolded on the fireground. Fire scene photographs of the various structures, including site and floor plan diagrams, were also provided; they are shown extensively in this book. Additionally, the tactical positions of the crew members and their actions, the placement of apparatus, and the advanced handlines and master streams used were also made available in the reports. Finally, the performance of radios and of the personal protective equipment (PPE) in use at the time was also noted.

All incidents were closely examined, with special attention given to the following:

- The information about the structures involved
- The conditions found upon and after arrival

- The strategy and tactics used
- The actions that occurred in the structure

The structural characteristics associated with the 17 buildings examined in the disorientation study varied widely. These structure fires included the Worcester, Massachusetts, Cold Storage Warehouse fire of 1999 (fig. 1–2) and the Phoenix Southwest Supermarket fire of 2001.

Fig. 1–2. The Worcester Cold Storage Warehouse fire. In 1999, with light smoke showing upon arrival, Worcester firefighters implemented a fast and aggressive interior attack into this large, enclosed structure involving a cold storage warehouse. However, during the operation, interior conditions quickly deteriorated from moderate to zero visibility, contributing to the disorientation and fatalities of six career firefighters. (Courtesy of Scott LaPrade.)

Other incidents included a wind-driven high-rise apartment fire in New York City; an auto repair shop fire in Chicago; a warehouse fire in Los Angeles; high-rise apartment fires in St. Louis and Houston; a small, two-story town house fire in Washington, DC; and a paper warehouse fire in Kansas City, Missouri.

Four fires that did not involve fatalities occurred at a large place of assembly, a three-story office building, a thrift store, and a converted single family residence, all in San Antonio, Texas. Other specific structure fires examined included a single-family residence in Crooksville, Ohio; a large auto salvage warehouse in Goldsboro, North Carolina; a massive multistory warehouse in Fall River, Massachusetts; and a grocery store in Fairlea, West Virginia.

Firefighter disorientation was also shown to strike departments regardless of size, type, or geographical location. Furthermore, and as an indication of the extreme degree of danger present during these specific types of fires, no category of firefighter was immune from the disorientation hazard, as it took the lives of individual firefighters regardless of rank, age, training, assignment, or level of experience. The vulnerability of firefighters who are taken by disorientation fires has not changed through the years, as indicated by the 2007 tragedy that claimed the lives of nine career firefighters at the Sofa Super Store Fire in Charleston, South Carolina. The loss of life at this incident illustrated in basic terms that anyone can be susceptible to disorientation.

Firefighter Disorientation

Firefighter disorientation can be defined as "the loss of direction due to the lack of vision in a structure fire" (Mora 2003). This simple definition accurately describes the complex and dangerous situation distressed firefighters experience while working in the interior of a structure fire. Commonly expressed by firefighters as "the inability to see your hand in front of your face," firefighter disorientation hazard specifically refers to a firefighter's sustained loss of vision. Once perceived to involve isolated cases of firefighters lost during heavy smoke conditions in large commercial structures, subsequent study revealed the magnitude of the problem to be considerably more significant and complex. It is now known that this problem requires a multipronged approach to effectively address each of the components that contribute to disorientation and structural firefighter fatalities. It is also clear that firefighter disorientation is such an intricate and prevalent event during interior firefighting operations that it should be considered the greatest and most persistent contributing factor leading to traumatic line-of-duty deaths in structure fires. Moreover, analysis of the traditional method of operation used at disorientation fires showed this method to be unsafe and ineffective. In light of this finding, a major operational shift is immediately required in the fire service if fatalities involving firefighter disorientation are to be prevented. In this effort, every active firefighter must be clearly informed of the hazard and must be trained, equipped, and completely prepared to manage it in the safest manner possible. In short, since this firefighting hazard is national in scope, the approach used to prevent disorientation fatalities must be institutionalized in every department in the country. This process begins by accurately recognizing, defining, and understanding the danger associated with various structures and conditions commonly encountered on the fireground, followed by an understanding of findings of the disorientation study.

With a firm grasp of critical and fundamental operational issues such as the use of wind-driven fire action plans, use of quicker developing hose evolutions, and knowledge of the hazards that lead to firefighter disorientation, a close review of past disorientation cases can be made. The review, conducted in chapter 13, first examines the effectiveness of the offensive strategies initially used during each disorientation incident, followed by the use of enclosed structure tactics developed for and dedicated to extremely dangerous enclosed structure fires.

References

Fahy, R. F., LeBlanc, P. R., and Molis, J. L. 2015. "Firefighter Fatalities in the United States—2014." http://www.nfpa.org/research/reports-and-statistics/the-fire-service/fatalities-and-injuries/firefighter-fatalities-in-the-united-states.

Mora, W. R. 2003. "U.S. Firefighter Disorientation Study 1979–2001." http://www.sustainable-design.ie/fire/USA-San-Antonio_Firefighter-Disorientation-Study_July-2003.pdf.

2

TERMS, RISK, AND ASSUMPTIONS

To effectively address the chronic and often fatal firefighter disorientation hazard, distinct terminology is needed. Understanding the meaning of these terms is important so that every firefighter is able to refer to and to communicate the risk associated with this major hazard in a simple and mutual language. The terms specifically refer to the types of blinding smoke, fire conditions, and structures commonly encountered and their associated degree of danger. Finally, dedicated and defined tactics and tasks necessary to safely and effectively manage the risk on the fireground are also provided and used. These new terms will likewise give firefighters the ability to understand the different types of disorientation that can occur during working fires and, therefore, how to avoid them. Other terms were developed to describe the manner in which risk is safely assessed and gauged and the manner in which safer tactics are initiated.

When attempting to determine the amount of danger associated with structures involved in fire, it is important to know that a building's structural design serves as a major indicator. In this regard, it has been determined that buildings incorporate five fundamental structural designs: opened structures, opened structures with a basement, enclosed structures, enclosed structures with a basement, and high-rise hallways, each of which is defined in this chapter.

Opened Structure Defined

When firefighter disorientation cases were studied (Mora 2003), several features pertaining to the incidents were closely examined, including the structure, the smoke, and fire conditions present upon and after arrival, and the action taken by arriving firefighters. One of many findings was that, although disorientation involving blinding smoke did occur in structures having an enclosed design, disorientation involving the blinding effects of fire could also initially occur in any structure, followed by zero-visibility conditions. The link to disorientation and the frequency of these specific events involved the presence or absence of windows or doors, which was in turn tied to the ability to promptly ventilate or safely evacuate the structure. In one case examined during the study's time span, a line-of-duty death attributed to the firefighter disorientation factor took place in Pensacola, Florida, in an unprotected (no operable sprinkler system) residential structure with several windows and doors. This incident, which occurred in a structure of moderate size with an opened structural design, underscored the fact that firefighter disorientation, whether initiated by smoke or fire, can happen in any structure. These particular structures were referred to as opened structures to emphasize the presence of windows or doors and for size-up purposes to accurately describe their exterior appearance. An *opened structure* is one that "has windows or doors of sufficient number and size to provide for prompt ventilation and emergency evacuation" (Mora 2003).

Opened structures, as seen in figures 2–1 and 2–2, typically do not have operable sprinkler systems, are one or two stories in height, are small to moderate in size, and are constructed on a concrete slab foundation. Therefore, opened structures have no involved basement level to enter or possibly to fall into. They can be of any type of occupancy or construction and may be occupied, unoccupied, or vacant. An opened structure also does not have burglar bars or boarded doors or windows.

2   Terms, Risk, and Assumptions

Fig. 2–1. The most common example of an opened structure is a single-story, single-family residence. Typically built on a concrete slab foundation, an opened structure does not have an operable sprinkler system, nor does it have burglar bars, boarded windows, or a basement.

Fig. 2–2. Opened structures can be found in any community, such as this older, two-story retail structure located in Annapolis, Maryland.

Any two-story structure, residential or retail, can also be considered opened, provided an adequate number of windows and doors are present to allow for prompt ventilation and emergency evacuation of firefighters from both levels. The assumption made here is that, if necessary, a firefighter trapped on the second floor could open a window, hang from the windowsill, drop to the ground, and survive. Examples of two-story residential and retail opened structures are given in figures 2–3, 2–4, 2–5, and 2–6.

Fig. 2–3. This is a two-story residential opened structure with an A-B view.

Fig. 2–4. In established neighborhoods, firefighters must learn to instinctively focus beyond shrubbery and fencing to determine if means of evacuation, such as windows and doors, are present.

Fig. 2–5. Regardless of the occupancy type, any two-story structure is considered "opened" as long as it is small to moderate in size and has adequate numbers of windows and doors to provide prompt ventilation and emergency evacuation of firefighters.

Fig. 2–6. This is a small retail store with an opened structural design.

In addition to doors, the availability and associated egress features found with various windows determine whether firefighters can survive should interior firefighting conditions deteriorate. Various windows are illustrated in figures 2–7, 2–8, and 2–9.

Fig. 2–7. For safety and survival, windows must be accessible, penetrable, and large enough to allow firefighters, wearing full personal protective equipment, to quickly execute an emergency evacuation. (Courtesy of Peter Matthews.)

Fig. 2–8. Windows provide an emergency means of egress. Depending on the situation encountered, firefighters may have time to open and climb out of a window. However, during rapidly changing conditions, they may need to suddenly break through a window and the window frame as safely and quickly as possible.

Fig. 2–9. Due to limited movement of window panes, certain types of windows do not allow for prompt evacuation of firefighters. This commercial structure has numerous metal-framed factory windows, which do not open outward far enough for firefighters to make a rapid evacuation.

Firefighters typically respond to structure fires within various types of districts such as commercial, industrial, mixed-use, or residential. A few examples of structures located within residential districts, where firefighters respond most often, are shown in figures 2–10, 2–11, 2–12, and 2–13.

Fig. 2–10. Most structures in residential districts include an opened structure design with ample means of egress.

Fig. 2–11. Single- and two-story structures are a significant part of residential districts.

Fig. 2–12. Entire neighborhoods are often comprised of small- to moderate-sized single- and two-story opened structures.

Fig. 2–13. Although dangerous during a fire, opened structures do allow for prompt ventilation and emergency evacuation of firefighters.

A two-story garden apartment building that may be large in overall size is considered an opened structure as long as it has an adequate number of readily penetrable windows and doors available from each subdivided unit (figs. 2–14 and 2–15). Opened structures are dangerous during a fire because of the risk of disorientation, but they also allow for fast firefighter evacuation.

Fig. 2–14. A two-story garden apartment with an opened structure design.

Fig. 2–15. Two-story garden apartments such as this and the one in figure 2–14 are considered opened structures. However, they can expose firefighters to multiple hazards including electrocution, natural gas explosion, flashover, backdraft, wind-driven fire, and the collapse of lightweight truss floors and roofs. Disorientation caused by exposure to flashover, wind-driven fire, collapse, and prolonged zero-visibility conditions is therefore a possibility.

Enclosed Structure Defined

One major finding uncovered during the course of Mora's (2003) disorientation study was that the design of a structure in conjunction with the absence of an operable sprinkler system had major ramifications for the safety and survival of firefighters during a quick and aggressive interior attack. The results showed that firefighter disorientation occurred in all of the 17 cases examined. In 100% of those cases, the involved structures had an enclosed structural design, and 88% did not have an operable sprinkler system. These specific structures are referred to as enclosed structures, which describes their exterior appearance. An *enclosed structure* has an absence of windows or doors of sufficient number and size to provide for prompt ventilation and emergency evacuation (Mora 2003).

In other words, the means of egress are secured, scarce, too high up from the floor to be accessible, or not large enough to allow for the prompt and safe evacuation of fully bunkered firefighters when conditions deteriorate. Effective ventilation is also very difficult or impossible to accomplish promptly during the course of an incident. Enclosed structures that are not protected by an operable sprinkler system should be considered extremely dangerous because firefighter fatalities repeatedly occur in such structures. A structure can be enclosed either by architectural design or by alteration after completion of construction. This can be done in a number of ways, including by enclosing preexisting windows or doors by use of such materials as burglar bars, wooden boards, metal sheeting, plywood sheeting, brick and mortar, cinder block, and gypsum board (figs. 2–16 and 2–17).

Fig. 2–16. Structures can be enclosed in various ways. The upper-floor windows of this four-story enclosed commercial structure have been covered with plywood sheeting, while burglar bars secure the first-floor windows. The D wall through the entire height of the structure is blank. In addition to the hazardous enclosed condition, the structure may not be equipped with an operable sprinkler system.

Fig. 2–17. The preexisting doors and windows of this structure have been enclosed by brick and mortar and by plywood sheeting.

In addition, structures having windows or doors covered by steel security bars or metal gates of any kind should be considered enclosed and extremely dangerous. Although the presence of security bars will not prevent ventilation from taking place, they will prevent an emergency evacuation should interior conditions rapidly deteriorate. During an interior attack, security bars may trap firefighters in the building if the point of entry becomes blocked by fire and/or by debris following a partial or complete collapse of the roof (figs. 2–18, 2–19, 2–20, and 2–21).

Fig. 2–18. Known to be involved in the fatality, injury, and narrow escapes of firefighters, unprotected enclosed structures of any size that have burglar bars or other similar types of security should be considered extremely dangerous.

Fig. 2–19. Note the burglar bars and cinder block used to enclose the windows in this vacant four-story structure. Although burglar bars will not prevent effective ventilation, they will prevent firefighters from achieving a prompt emergency evacuation should interior conditions rapidly deteriorate.

Fig. 2–20. Nationally, countless structures of various occupancies intentionally incorporate an enclosed architectural design.

Fig. 2–21. Any structure, regardless of the size, age, configuration, or type of construction, can have an enclosed design and include various occupancies such as theaters, restaurants, museums, and offices.

Degrees of Danger

Firefighters frequently work in dangerous places, called immediately-dangerous-to-life-and-health (IDLH) environments. However, because of effective, standardized, and integrated risk management efforts, not all dangerous environments ultimately take the lives of firefighters. In fact, in virtually every structure fire operation in the country, structure fires are extinguished safely in traditional ways. These operations may use offensive strategies, with firefighters advancing handlines into and operating on the interior of the structure, or defensive strategies, with firefighters operating master streams from the exterior of the structure. And, in virtually every operation, every firefighter does go home after a tour of duty. Therefore, in reality, most firefighters do safely and routinely manage dangerous environments by the use of traditional strategy and tactics, the latest safety technology, and the Incident Command System (ICS). On the other hand, to understand why firefighters continue to die traumatically during structure fires in spite of this, it is necessary to first speak of the conditions on the fireground in terms of "degrees of danger." This understanding will then allow firefighters to manage the previously unrecognized risk more safely and effectively, thus preventing future fatalities.

Although exposure to any one of a number of hazards encountered at the scene of a structure fire may take the life of a firefighter, the word "dangerous" is used to describe the hazardous nature of opened structure fires, given that firefighting in these structures is only occasionally fatal. "Extremely dangerous" is used to describe enclosed structure fires because firefighting operations in these structures take the lives of firefighters at a disproportionate rate.

In any structure, opened or enclosed, initiating interior attacks can be fatal to firefighters. Therefore, an understanding of a few fundamental assumptions about opened and enclosed structure fire is necessary.

Opened structure assumptions

Opened structures are small to moderate in size, built on concrete slab foundations, and exist in various occupancies and configurations in every community. They are identified visually by their open appearance, but firefighters must also understand why they are considered dangerous but not extremely dangerous.

The worst-case scenario assumption for opened structures

When considering interior operations in an opened structure, a worst-case scenario approach must be taken when attempting to safely manage this type of structure fire.

This assumption recognizes that the incident may involve a working fire in a small- to moderate-size structure with heavy fuel loading, requiring firefighters to advance handlines into the building to knock down and completely extinguish the fire. Opened structures as defined are not equipped with operable sprinkler systems, and interior firefighters may be exposed to life-threatening hazards that may include but are not limited to prolonged zero-visibility conditions, flashover, backdraft, wind-driven fire, and collapse of a roof or second floor.

Enclosed structure assumptions

Enclosed structures exist in different sizes, including small, moderate, and large. They also exist in various occupancies and configurations. Enclosed structures can be occupied, unoccupied, or vacant during the course of a fire. However, in order to identify, categorize, and understand why they are so dangerous, an understanding of certain assumptions associated with enclosed structures is necessary.

The worst-case scenario assumption for enclosed structures

As experience involving firefighter fatalities and use of fast and aggressive interior attacks has repeatedly shown, firefighters must always take a worst-case scenario approach when safely managing interior operations at enclosed structure fires.

This assumption recognizes that the incident may involve a working fire in a small- to moderate-size enclosed structure with a heavy fuel load, requiring firefighters to advance handlines into the building to knock down and completely extinguish the fire. Additionally, life-threatening hazards that may expose these interior firefighters may include but are not limited to prolonged zero-visibility conditions, flashover, backdraft, wind-driven fire, and collapse of a fire-weakened floor or roof.

This assumption recognizes that the incident may involve a deep-seated working fire in a large enclosed structure with heavy fuel loading and complex interior floor plan arrangements, requiring firefighters to advance handlines and or portable monitors into the building to knock down and completely extinguish the fire. Additionally, life-threatening hazards that may expose advancing or evacuating firefighters may include but are not limited to prolonged zero-visibility conditions, flashover, backdraft, wind-driven fire, and collapse of a fire-weakened floor or roof.

The large structure assumption

When determining whether a structure is an opened or enclosed structure, there comes a point when considering the size or area of a structure where, because of increasing risk, it can no longer be included in the opened structure category. Recall that opened structures are small to moderate in size and allow prompt ventilation and emergency evacuation to take place. Therefore, for classification purposes, any large structure defined as one measuring approximately 100 ft. × 100 ft. or greater in size, regardless of the number of windows or doors present along the perimeter, is considered enclosed. Structures of this size are considered enclosed because when assuming a worst-case scenario, it would be very difficult to quickly ventilate and evacuate a structure of this or greater size in prolonged zero-visibility conditions from deep interior positions (fig. 2–22). It is also very important to be aware that structure fires of this size are associated with multiple firefighter fatalities at single fire events. Due to their size, large residential structures also fall into the large enclosed structure category and, if not protected by an operable sprinkler system, are extremely dangerous (fig. 2–23).

Fig. 2–22. Due to sheer size, prompt ventilation would be difficult to achieve during a working fire at this large enclosed structure. The resulting combination of prolonged zero-visibility conditions, obstacles, and long distances to points of entry would also prevent safe evacuation prior to depleting all available air in commonly used 30-minute rated self-contained breathing apparatus (SCBA).

Fig. 2–23. Operations involving heavy smoke conditions in larger structures, including larger residential occupancies, often considered by firefighters as "only dangerous," make ventilation and evacuation more difficult during deep interior attacks. Therefore, any large structure not protected by an operable sprinkler system regardless of the occupancy type should be considered extremely dangerous during the course of a working fire.

The tall structure assumption

When determining whether a structure is an opened or enclosed structure, there comes a point when considering the height of a structure where it can no longer be included in the opened structure category regardless of the number of windows present along the perimeter. Any structure having a height of three stories or greater than 20 ft., or 10 ft. per floor, should be considered enclosed, because of the danger that firefighters might not be able to promptly ventilate or safely evacuate a fire in blinding smoke conditions when it takes place on an upper floor. This assumption recognizes that a fall from a third-floor or higher window may be fatal. Therefore, any structure of three stories or greater in height is classified as an enclosed structure due to the dangerous height of the structure (figs. 2–24, 2–25, and 2–26).

Fig. 2–24. Assuming a worst-case scenario, any building of three or more stories is considered an enclosed structure having the same degree of danger associated with other enclosed structures.

Fig. 2–25. Regardless of the number of windows present along the perimeter, any structure having three or more stories or any structure that measures approximately 100 ft. × 100 ft. or greater is considered large and enclosed.

Fig. 2–26. These structures are all considered large enclosed structures.

The fall height criteria

The report "Guidelines for Field Triage of Injured Patients: Recommendations of the National Expert Panel on Field Triage" (Sasser, Hunt, Sullivent, et. al. 2009) provides a rationale for retention of the current fall height criteria. Although referencing traumatic injury to citizens, it can also pertain to trapped firefighters in recognition of the risk associated with falls from structures of three stories or more (or greater than 20 ft. high). According to the report:

> The extent of injury from a fall depends on characteristics of the person, the distance fallen, the landing surface, and the position at impact. A 5-month prospective study of trauma team activations at a level I trauma center indicated that 9.4% of victims who fell >20 ft suffered injuries serious enough to require ICU [intensive care unit] admission or immediate operating room intervention.

The report goes on to state:

> In reaching its conclusion, the Panel noted that the fall height criterion for adults of >20 ft. has been a component of the Decision Scheme since 1986 and is familiar to prehospital providers and their medical directors. In addition, the Panel took note of the established relationship between increase fall height and increase risk for head injury, death, ICU admission, and the need for operating room care. The Panel concluded that in absence of new evidence that establishes a definitive height for this criterion or that supports changing or eliminating the criterion for falls of >20 ft. for adults (with 10 ft. equivalent to one story of a building), this criterion should be retained, and adult patients who fall >20 ft. should be transported to the closest appropriate trauma center for evaluation.

Therefore, and while establishing degrees of danger and incorporating this guideline on the fireground, opened structures (which can be a maximum of two stories in height or 20 ft. total) are considered dangerous but not extremely dangerous. This is because firefighters can ventilate and evacuate the structure with relative safety, even though they may be forced to initially hang from a windowsill and drop from a second-story window (10 ft.) to escape. However, this is not the case with unprotected structures with heights of three stories or greater. Structures of three or more floors that are not equipped with an operable sprinkler system and/or required standpipe system are to be considered enclosed and extremely dangerous, because sprinkler heads will not be controlling fires at an early stage and a standpipe will not be available for use by firefighters. Additionally, a fall of greater than 20 ft. is considered a potentially fatal fall for any firefighter who needs to make an emergency evacuation (figs. 2–27 and 2–28).

Documented experiences of members of fire departments in the state of New York underscore this extreme danger. During a working structure fire, two fully bunkered New York City Fire Department (FDNY) firefighters were forced to bail out, headfirst, from a second-story window due to a flashover, and both survived. However, during an incident in 2009, three Yonkers Fire Department firefighters were operating at a working fire in a three-story residence. The crew was on the top floor of the structure when fire conditions changed rapidly. The three firefighters were forced to jump from the structure to the ground. One of these firefighters received fatal injuries in the fall (U.S. Department of Homeland Security 2010).

In another case, in 2005, six FDNY firefighters who were disoriented and trapped by fire were forced to jump from fourth-floor windows at an illegally subdivided and unprotected apartment building. Although four survived the fall, two firefighters did not (NIOSH 2005).

Fig. 2–27. Three-story structures of any occupancy are considered extremely dangerous enclosed structures, because a fall of 20 ft. or more may be fatal.

Fig. 2–28. Structures can be enclosed by architectural design or by alteration after construction. Cinder block and plywood sheeting were used to enclose the windows of this five-story structure, thereby eliminating access to the fire escapes. By simple observation, firefighters can easily determine that this structure is large, tall, and enclosed.

The unprotected and protected enclosed structure assumptions

The danger associated with an inability to promptly and safely evacuate an enclosed structure is influenced primarily by the fixed fire protection available. In 88% of the disorientation cases examined, there was no sprinkler system in the structure or the system did not operate (Mora 2003). The unprotected enclosed structures were therefore extremely dangerous. High-rise fires at the Deutsche Bank building in New York City (fig. 2–29) and the One Meridian Plaza building in Philadelphia are prime examples of the increased level of danger firefighters may encounter when responses are made to enclosed high-rise structure fires that are fully or partially unprotected.

Fig. 2–29. In 2006, two FDNY firefighters became disoriented in blinding smoke and depleted their air supply while attempting to exit an upper floor at the vacant Deutsche Bank building. At the time of the fire, plywood sheets were used in the interior, creating a maze-like situation. The firefighters were also at a great disadvantage during the incident because the sprinkler and standpipe systems were disconnected.

The 1991 One Meridian Plaza Incident

More than 20 years ago, the U.S. Fire Administration (USFA), in Technical Report 049, noted the danger associated with high-rise structures that are unprotected by automatic sprinkler systems and operable standpipe systems (Routley, Jennings, and Chubb 1991). According to the report involving a 1991 Philadelphia, Pennsylvania, incident: "Three firefighters from engine company 11 died on the 28th floor when they became disoriented and ran out of air in their SCBAs [self-contained breathing apparatus]." The three firefighters who died were attempting to ventilate the center stair tower. They radioed a request for help stating that they were on the 30th floor. After extensive search and rescue efforts, their bodies were later found on the 28th floor. They had exhausted all their air supply and could not escape to reach fresh air. An eight member search team became disoriented and ran out of air in the mechanical area on the 38th floor, while trying to find an exit to the roof. According to the report, "The fire was eventually stopped when it reached the fully sprinkled 30th floor. Ten sprinkler heads activated at different points of fire penetration."

In reference to the safety issue during high-rise building fires, the report states:

> When things go wrong on a scale as large as One Meridian Plaza, safety becomes an overriding concern. Firefighters were continually confronted with unusual danger caused by multiple system failures during this incident. The deaths of the three firefighters and the critical situation faced by the rescue team that was searching for them are clear evidence of the danger level and the difficulties of managing operations in a dark, smoke-filled highrise building.

The 1988 First Interstate Tower Incident

> On Wednesday, May 4, and continuing into May 5, 1988, the Los Angeles City Fire Department responded to and extinguished the most challenging and difficult high-rise fire in the city's history. The fire, which originated on the 12th floor, destroyed four floors and damaged a fifth floor of the modern 62-story First Interstate Bank building in downtown Los Angeles. The fire claimed 1 life, injured 35 occupants and 14 fire personnel, and resulted in a property loss of over 50 million dollars. Although it was built in 1973, one year before a high-rise sprinkler ordinance went into effect, and had protection only in the basement, garage, and underground pedestrian tunnel, a complete automatic sprinkler system costing 3.5 million was being installed in the building at the time of the fire. However, a decision had been made to activate the system only on completion of the entire project, when connections would be made to the fire alarm system, so the valves controlling the sprinklers on completed floors were closed. . . . It demonstrated the absolute need for automatic sprinklers to provide protection for tall buildings. (Routley 1988)

Through experience and observation over time, enclosed structures that are not equipped with an operable sprinkler system have come to be considered extremely dangerous during a fire (fig. 2–30).

Preventing Firefighter Disorientation

Fig. 2–30. All enclosed structures, including enclosed high-rise structures, without the protection of an operable sprinkler system should be considered extremely dangerous. During the course of a working fire, sustained, blinding smoke conditions and other life-threatening hazards should always be anticipated. Of the structures shown here, only those not equipped with an operable sprinkler and (if required) a standpipe system should be considered extremely dangerous.

Although a structure may be enclosed, when protected by an operable sprinkler system or (if required) an operable standpipe system, the structure is considered "only" dangerous rather than extremely dangerous. In spite of this designation, it is important for firefighters to further understand that dangerous means that firefighter fatalities can occur, but at a lower rate. In this regard, it is very important to stress that even when protected by an operable sprinkler system, operations in an enclosed structure must still be considered potentially fatal because, as history has shown, firefighters have lost their lives in enclosed structure fires even when protected by an operable sprinkler system. Precisely such an incident occurred in 1987 in Fall River, Massachusetts. The following account was provided by Lieutenant Clifford Clement of the Fall River Fire Department, who was an active participant in the incident.

The 1987 Fall River Incident

In 1987, firefighters of the Fall River (Massachusetts) Fire Department found heavy smoke showing from the loading dock at this large, three-story, enclosed structure involving a dye-processing facility (fig. 2–31). After initiating an aggressive interior attack on the fire by advancing through the loading dock door, they began to run low on air. As a result, the first-arriving company, comprised of an officer and two firefighters, began to retrace their handline out of the structure. However, the handlines became entangled in the building, and two of the three firefighters became disoriented by the blinding smoke and confused by the entangled hose. One of the crew members was able to exit the structure alone, and another arriving company attempted to rescue the two who were still in the building: an officer and a firefighter. The rescue team was able to bring one firefighter out of the building; however, the disoriented officer became separated from his crew and unknowingly followed another handline away from the point of entry toward the area near the first door (see fig. 2–31). In spite of an aggressive rescue attempt, in which rescuers also risked their lives, the officer exhausted his air supply and died of his injuries before his location was determined and rescue was possible.

This large enclosed structure incident presented numerous hazards to arriving firefighters, such as preexisting windows that had been filled in with cinder block; prolonged zero-visibility smoke conditions generated by a burning commercial oven; suspended and wet dye that had coated the masks, hose, hand lights, and firefighting tools used during the attack; and the enormous size of the structure. It is important to note that although the seat of the fire was located only 20 ft. from the point of entry and the structure was protected by an operable sprinkler system, the hazards encountered collectively worked against the firefighters and ultimately contributed to firefighter disorientation, the fatality of one officer, and the narrow escape of other firefighters involved in the operation.

Fig. 2–31. Fall River (Massachusetts) Fire Department firefighters made an interior attack on a working fire at this large, protected enclosed structure. Even when enclosed structures are protected by operable sprinkler systems, they must nonetheless be considered dangerous as firefighter fatalities during interior operations can take place. (Courtesy of Lieutenant Clifford Clement, Retired.)

Since the Fall River incident, other fatal disorientation fires in large, unprotected, enclosed commercial structures involving grocery stores in Phoenix, Arizona, and Fairlea, West Virginia, as well as warehouses in Los Angeles and Salisbury, North Carolina, illustrate the continuing degree of danger firefighters can be exposed to when there is no operable sprinkler system. Figures 2–32, 2–33, 2–34, and 2–35 provide more information about enclosed structures of various age and occupancies.

Fig. 2–32. This preexisting window has been filled in with matching brick and mortar, thus enclosing the structure. However, it is also fully protected by an automatic sprinkler system making it only "dangerous."

Fig. 2–33. Enclosed structures can be of any age, height, or type of construction. The ornate structure on the right was built in 1865; the taller, more modern high-rise of fire-resistive construction on the left was built more than 100 years later. If not protected by an operable sprinkler and standpipe system, both would be extremely dangerous during the course of a working fire.

Preventing Firefighter Disorientation

Fig. 2–34. Subway stations and tunnels are considered enclosed spaces. Due to the inability to promptly ventilate, fires may produce heavy and prolonged zero-visibility conditions.

Fig. 2–35. Shopping malls are considered dangerous large enclosed structures even when protected by an operable sprinkler system.

Using the Terms "Opened" or "Enclosed" for Added Safety

The terms "residential" and "commercial" are often used in broad reference to structures of different occupancy types. Firefighters also use the terms as a means of informally expressing the danger or difficulty of entering and attacking these types of structures and of the life safety hazard potential. In practice, however, it may not be entirely safe to use the terms in this context since a residential structure may not always have an opened design, and a commercial structure may not always have an enclosed design (fig. 2–36).

Fig. 2–36. Opened structures are not all residential, as commercial structures can have an opened design. This structure, measuring 60 ft. × 40 ft. and used as a media production company, has an opened design and is considered only "dangerous," not "extremely dangerous."

In other words, by using the occupancy as a means to exclusively define the danger, firefighters may fail to accurately address the disorientation hazard and therefore adversely and seriously affect safety. As an example, a residential structure, when built on a concrete slab foundation, can have an opened design when adequate windows and doors exist and be considered only "dangerous," but residential structures can also have an enclosed design if, for example, burglar bars, boards, or plywood sheeting have been used to secure all windows and doors. In this case, the residential structure would be extremely dangerous. Therefore, when needed, it may be safer to describe a residential or commercial structure according to the structural design by using the phrases *opened residential structure, enclosed residential structure, opened commercial structure,* or *enclosed commercial structure* to more accurately describe the type of structure encountered and potential risk during the initial size-up.

Other examples of residential structures that fall into the enclosed structure category can include a multistoried structure that is three or more stories in height and serving as a hotel, apartment, condominium, townhouse, or a row house. Again, the degree of danger associated with a residential or commercial structure is ultimately determined not by the occupancy type, but whether or not the structure is enclosed and protected by an operable sprinkler or standpipe system (fig. 2–37). If, for example, a sprinkler valve on the 10th floor of a residential apartment high-rise structure was closed, the area and occupants it was intended to serve would be unprotected, and thus the structure would be considered extremely dangerous to civilians and firefighters alike should a working fire break out.

In summary, and from the firefighter safety perspective, the criteria determining whether a structure is considered opened or enclosed, and therefore dangerous or extremely dangerous, partially depend on the ability of firefighters to promptly ventilate and safely evacuate. However, the structure must have an adequate number of windows or doors in relation to the size of the structure, and the building height must be considered. Therefore, regardless of the number of windows present in the large structure in figure 2–38, because it has a dangerous height of more than three stories, it is considered enclosed. However, the degree of danger is ultimately determined by presence or absence of an operable sprinkler system.

Fig. 2–37. Although greater than three stories in height, because the high-rise hotel in the center of the photo is fully sprinkled, it is considered only "dangerous."

Fig. 2–38. Since the structure is fully protected by an operable sprinkler system, the degree of danger associated with this three-story apartment building is only "dangerous."

References

Mora, W. R. 2003. "U.S. Firefighter Disorientation Study 1979–2001." http://www.sustainable-design.ie/fire/USA-San-Antonio_Firefighter-Disorientation-Study_July-2003.pdf.

National Institute for Occupational Safety and Health (NIOSH). 2005. "Career Lieutenant and Career Fire Fighter Die and Four Career Fire Fighters Are Seriously Injured During a Three Alarm Apartment Fire—New York." http://www.cdc.gov/niosh/fire/reports/face200503.html.

Routley, J. G. 1988. "Interstate Bank Building Fire, Los Angeles, California" (USFA-TR-022/May 1988). http://www.usfa.fema.gov/downloads/pdf/publications/tr-022.pdf.

Routley, J. G., Jennings, C., and Chubb, M. 1991. "Highrise Office Building Fire: One Meridian Plaza, Philiadelphia, Pennsylvania" (USFA-TR-049/February 1991). http://www.usfa.fema.gov/downloads/pdf/publications/TR-049.pdf.

Sasser, S. M., Hunt, R. C., Sullivent, E. E., et al. 2009. "Guidelines for Field Triage of Injured Patients: Recommendations of the National Expert Panel on Field Triage." (CDC, MMWR Recommendations and Reports/January 2009). Vol. 58. 25–26. http://www.cdc.gov/mmwr/preview/mmwrhtml/rr5801a1.htm.

U.S. Department of Homeland Security, Federal Emergency Management Agency, U.S. Fire Administration, National Fire Data Center, National Fallen Firefighters Foundation. 2010. "U.S. Fire Administration Firefighter Fatalities in the United States in 2009." http://www.usfa.fema.gov/downloads/pdf/publications/ff_fat09.pdf.

3

TYPES OF ENCLOSED STRUCTURES

Enclosed Structure Characteristics

An enclosed structure is one of two general types of structures—opened or enclosed. However, the term *enclosed structure* is also used to refer to one of the enclosed structure subclasses. A key characteristic in identifying these structures is that enclosed structures encompass a wide variety of uncommon features. For example, it can be of any *type of occupancy*, including but not limited to a single-family residence, a multifamily residence, a restaurant, a church, a grocery store, an auto repair shop, a pawn shop, a bowling alley, a furniture store, a nightclub, a theater, a warehouse, a massive shopping mall, or a high-rise building.

Structures having an enclosed design can also be structures that were converted at one or more points during their life span. For example, a structure may have originally been built as a movie theater but later converted and now used as a thrift store. A small convenience store may have been converted into a single-family dwelling or a grocery products warehouse could have been converted into a paper warehouse. A grocery store may have undergone expansion to ultimately be used as a furniture store.

An enclosed structure can also be of any *type of construction*. The structures examined during the disorientation study (Mora 2003) varied and included five that were of fire-resistive construction, two of heavy timber construction, seven of unprotected noncombustible construction, and three of protected wood frame construction.

The *size* of an enclosed structure can also vary widely. In the study, the sizes ranged from a townhouse unit measuring 19 ft. in width and 33 ft. in length to a paper warehouse measuring 500 ft. in width, 600 ft. in length, and 25 ft. in height.

Additionally, the *age* of an enclosed structure can also vary. The study examined structures that included a 20-year-old, single-family dwelling; a thrift store built in the 1940s; and a 98-year old warehouse (Mora 2003). The enclosed structure can be in *any condition*, from structures that are well maintained to those that are allowed to become dilapidated and unsafe.

Another feature of enclosed structures is that the *configuration* can vary. For example, an enclosed structure can be square or rectangular in shape and exist as a stand-alone structure constructed in the middle of a large parking lot. The structure can also be oval, irregular, or have an L or H shape (figs. 3–1 and 3–2). Therefore, enclosed structures can have any of several different structural characteristics; however, the single and most important structural similarity a firefighter must be aware of is that all of these structures have an enclosed design, and because structures and spaces with enclosed designs repeatedly claim the lives of firefighters during the course of working fires, they should be considered extremely dangerous.

3   Types of Enclosed Structures

Fig. 3–1. New York City's Solomon Guggenheim Museum is essentially a large enclosed structure with a spiral configuration.

Fig. 3–2. New York's Madison Square Garden has a round, enclosed design.

Opened Structure with a Basement

An *opened structure with a basement* has readily penetrable windows and doors of sufficient number and size to provide for prompt ventilation and emergency evacuation from grade but not from the basement level. These structures are typically small to moderate in size; can be of any age, type of construction, or type of occupancy; and have a partial or full basement (figs. 3–3 and 3–4).

Fig. 3–3. Note the walkout basement door and the flight of stairs leading to a side door on the D side of this two-story opened structure with a full basement.

Fig. 3–4. For security purposes, these particular basement doors have been covered with steel grills and would require forcible entry prior to initiating calculated basement operations.

3 Types of Enclosed Structures

Every firefighter must realize that a fire in a basement can be extremely dangerous, and if the wrong action is inadvertently taken, serious injury or fatality can occur. It is fundamentally important for firefighters to also understand that basements can be located in many different regions of the country and in different types of occupancies of different sizes. In keeping with the opened/enclosed structure method of risk assessment, it is also important to understand that they exist below structures that may have either an opened or enclosed structural design.

The firefighter who understands the degree of danger associated with an opened structure with a basement and uses this knowledge regularly during the initial size-up process will begin to significantly enhance safety for the entire operation.

The initial exterior appearance of an opened structure with a basement can indeed be fatally deceiving. For example, an innocent-looking, small residential structure can in fact be extremely dangerous if it has an enclosed space in the form of a basement that happens to be out of view (figs. 3–5 and 3–6). Additionally, firefighters should keep in mind that a room within a structure can also be enclosed by design or enclosed after construction. Furthermore, material used to enclose a structure can be as simple as a sheet of plastic. In 1995, a thin plastic panel was used to cover the interior surface of a window in a residence in a below-grade family room in Pittsburgh, Pennsylvania. The room filled with blinding smoke, which took the lives of three firefighters who became disoriented and ran out of air attempting to evacuate. The small residence measured only 20 ft. in width and 33 ft. in depth but had two enclosed spaces consisting of a lower-level family room constructed above a partially finished basement (Routley 1995).

Preventing Firefighter Disorientation

Fig. 3–5. The view of the A side of the Bricelyn Street residence gave arriving Pittsburgh firefighters no indication of a lower level or the presence of a basement. (Courtesy of U.S. Fire Administration [USFA].)

Fig. 3–6. This window in the lower-level family room was covered by a smooth plastic panel, making it difficult in heavy smoke for three disoriented firefighters with gloved hands to know that the window and means of egress existed. (Courtesy of USFA.)

When viewed only from the A side of the structure during a quick, initial size-up, opened structures with basements can often have the appearance of a safer, opened structure built on a concrete slab foundation. This is a common and dangerous situation for the untrained firefighter to be confronted with due to possible serious consequences, yet it takes place daily.

When the problem is viewed nationally, overwhelming numbers of case studies and observations have revealed that simply because of the greater use of an enclosed design, structural fires in certain areas of the country are more dangerous than in others. In this respect, firefighters serving in roughly the central and northeastern regions of the country must understand that this massive area is home to the greatest concentration of enclosed structures of all types, the most common of which include opened structures with basements in residential occupancies.

Opened structures with basements can be encountered as single-family dwellings; duplexes; and multifamily homes including row houses, townhouses, and garden apartments. Since they may involve a life safety hazard, arriving firefighters may tend to focus their attention entirely on the need to conduct a primary search while overlooking the signs of an extremely dangerous basement fire. On a larger scale, failure to initially recognize the entire response area as one having a heavy concentration of extremely dangerous enclosed structures, of all varieties, may ultimately result in tragedy during a working fire, specifically in an opened structure with a basement.

Common signs of a basement

Firefighters may be able to determine that a structure has a basement through the initial size-up process alone, but on other occasions, there may be absolutely no visual indication that a basement is present. Some common signs or ways first-arriving firefighters may be able to determine if a structure has been constructed with a basement include the following:

Conducting a 360-degree walk-around. Since it is possible that there may be no indication from a view of the A side alone that a structure has a basement, it is critical that a trained firefighter, with a thermal imaging camera if possible, conduct a complete 360-degree walk-around of the structure. Information sought, and which should be immediately reported, includes signs that may indicate that the structure has a basement, such as basement windows and basement doors, or whether fire or smoke is showing from the basement, as seen in figures 3–7 and 3–8.

Fig. 3–7. Obvious signs of a basement include descending exterior stairs and basement doors.

Fig. 3–8. An inspection of all sides of a structure may provide signs of a basement, which may include a drop-off along the foundation line, basement windows, or a flight of stairs leading to the rear door as seen on the C side of the Bricelyn Street residence. (Courtesy of USFA.)

Visible basement windows or doors. When plants, shrubbery, parked vehicles, or snow do not obstruct the view along the A, B, C, or D sides of a structure, basement windows or doors may be clearly visible to arriving firefighters. Otherwise, they may be blocked from view (fig. 3–9).

Fig. 3–9. Ground cover has partially covered the barred basement windows on both sides of the front door of this residence.

A drop-off along the foundation line. In many cases, a sudden drop-off along one side of a structure serves as a strong indicator of the existence of a basement.

A slope in the terrain along the foundation line. Builders will often make use of hilly terrain by constructing a structure into the side of a hill. The result is often a structure having a sloping terrain along the foundation line of the structure. These structures will typically have a basement (fig. 3–10).

Fig. 3–10. Basement windows and a slope of the terrain along the foundation line are indications that a structure has a basement.

Steps leading to the front, side, or rear door. According to the building industry, due to basement excavation guidelines, steps are often needed to reach a structure's raised first floor after completion for several reasons. First, the basement walls must be built to a height that will accommodate heating and air conditioning ductwork in the basement's ceiling space. Second, the basement hole has to be excavated so that it is deep enough to be situated below the frost line and, at the same time, not so deep as to cause water runoff into the basement. These combined requirements cause the elevation of the first floor to be slightly above grade level, which calls for use of stairs in an opened or enclosed structure with a basement (fig. 3–11). Therefore, in many cases, steps leading to a structure's front, rear, or side door is a strong indicator of a basement.

Fig. 3–11. A flight of steps leading to the front, side, or rear door indicate that the structure may have a basement. These are extremely dangerous opened residential structures with basements.

Asking occupants who have evacuated the structure. Accurate information quickly obtained during the early stages of an incident is very valuable. Occupants of residences or managers and employees of commercial structures who have evacuated the structure are highly reliable sources of information. They are to be sought out and questioned immediately on arrival to determine if the structure has a basement and whether they know if the basement is involved in the fire. They may also be able to tell you where the interior basement door is, which can be accessed when conditions are relatively safe and when the structural integrity of the first floor has been determined.

Referring to prefire survey plans. When it is difficult to quickly determine if a structure has a basement, referring to prefire surveys should indicate whether the structure was constructed with a partial, full, or walkout basement. When basements are present and involved, companies that are en route or on the scene should be immediately advised.

Area familiarity. One of the first safety issues a newly promoted officer or newly assigned firefighter should determine when assigned to a firehouse is whether basements can be found in the first-due area and within the regular alarm response district. In addition, and since their very lives depend on it, every firefighter must be familiar with the location of all enclosed structures in their response area, specifically structures with basements. This can be done during prefire surveys or by asking at an appropriate time during the course of emergency or nonemergency responses. Building owners may also allow a crew to enter the basement to look at the floor-to-ceiling assembly to determine if lightweight wooden or steel trusses or engineered lumber (all of which quickly fail after exposure to fire) was used or if thicker and stronger dimensional lumber was used in construction. Again, and on a larger scale, firefighters who respond to structure fires in the central and northeast regions of the country must realize that because they live and work in an area of the country having the highest concentration of enclosed structures and spaces involving basements, they are inherently at greatest risk.

Safely cutting an inspection hole into the first floor. When it is absolutely impossible to determine by any other means whether a structure has a basement, as a last resort, firefighters should determine whether a basement exists by cutting an inspection hole through the first floor from a safe exterior position.

Enclosed Structure with a Basement

An *enclosed structure with a basement* has an absence of windows and doors of sufficient number and size to provide for prompt ventilation and emergency evacuation from either grade, above grade, or basement levels. Figures 3–12, 3–13, 3–14, and 3–15 provide examples of features associated with an enclosed structure with a basement involving a vacant commercial structure.

Preventing Firefighter Disorientation

Fig. 3–12. Note the slight angle along the foundation line on the B side of this single-story enclosed structure with a full basement.

Fig. 3–13. Note the various angles along the foundation line on the A and D sides of this vacant commercial structure.

Fig. 3–14. Heavy security bars covering the front door and basement levels would hamper a rapid evacuation.

Fig. 3–15. Note the burglar bars on the windows and the angle along the foundation line on the D side, as well as how plywood sheeting at the D corner window is also covered by security bars.

Reduced Situational Awareness During Basement Fires

The conditions encountered at the scene of many structure fires are often complex. For example, a town home fire in Washington, DC, was the site of a double firefighter fatality incident. The complex in which the town home was located consisted of some buildings that were constructed with basements and others that were not. One common situation that proved to be critical in this fire outcome was that the command officer positioned on the A side did not know that the structure had a basement, although firefighters positioned on the C side did (NIOSH 1999a).

This reduced level of awareness and/or communication pertaining to the existence of a basement also existed during the Bricelyn Street incident in Pittsburgh and Seattle's Mary Pang fire, which involved a large, unprotected, enclosed commercial structure. It is certain that had command been aware of the presence of a basement during these fires and of the serious implications, the outcomes would have been much different.

However, it is interesting to note that during other fatal basement fires, specifically, the Ebenezer Baptist Church fire in Pittsburgh, the commercial structure fire in Breckenridge, Pennsylvania, and the 2007 millwork warehouse fire in Salisbury, North Carolina, command was aware that the structures had involved basements and that interior attacks, including advancing across first floors over involved basements, had been initiated. This may suggest that because of basement familiarity, the presumed strength of the type of construction used, the amount of time the fire may have been burning, and past positive outcomes produced by the use of fast and aggressive interior attacks, the risk associated with basement fires at the time may have been considered only "dangerous" and not "extremely dangerous"—having a higher probability of causing serious injuring or multiple firefighter fatalities. Since enclosed structures are associated with multiple life-threatening hazards, an assumption that universally considers fire in any enclosed structure extremely dangerous is needed in the fire service. This includes fires in enclosed spaces such as basements.

If smoke or fire from a basement is detected, it is absolutely critical for everyone en route to and on the fireground to be advised immediately by radio or upon arrival. Additionally, no one should be allowed to enter the structure at the outset or thereafter if the first floor has been determined to be unstable. Although this may seem an overreaction to an everyday occurrence, numerous cases have clearly shown that this degree of caution is not only needed but should also be standard policy in every department in the country. For safety, the assumption must also be made that the structure's first-floor level is about to collapse until proven otherwise. Firefighters must also bear in mind that although a lightweight wooden truss will collapse suddenly at any time after the trusses have been exposed to fire, any wooden beam of any dimension supporting the floor will also eventually burn through and cause the floor and furnishings to collapse when exposed to fire for a sufficient amount of time. And that time may transpire shortly after firefighter arrival. Since a traditional, quick, and aggressive interior attack is not always effective in these types of structures and, in fact, repeatedly results in tragic outcomes, risk management, focused on specific hazards before the fire ever occurs, must become an integral part of fireground operations. This involves using safe enclosed structure tactics, including those for enclosed spaces such as basements, and conducting primary searches in such a way that specifically avoids the multiple dangers associated with these high-risk, high-frequency types of structure fires linked to firefighter fatalities (fig. 3–16).

Fig. 3–16. Due to the stronger fire-resistive construction in many structures of various heights, the collapse potential of the first floor into the basement during a fire is greatly diminished. However, entanglement and disorientation due to prolonged zero-visibility conditions, flashover, and backdrafts must still be considered major concerns.

Opened and Enclosed Structure Combination

An opened and enclosed structure combination consists of a mixture of the opened and enclosed structure types. The combination type of structure may have adequate means of egress in one area and few or no means of egress in another, as shown in figure 3–17. When involved in fire and not protected by an operable sprinkler system, the enclosed area would be considered extremely dangerous while the opened area would be viewed as dangerous.

Fig. 3–17. This is an example of an opened and enclosed structure combination. While the first floor is completely enclosed, the second floor has an opened design. The degree of danger will therefore be mixed; extremely dangerous on the first floor, dangerous on the second.

Enclosed Spaces

This broad subcategory encompasses spaces of differing configurations, sizes, and functions, such as high-rise hallways, basements, attic spaces, ceiling spaces, stairwells, and stairways. As in the case of an enclosed structure, due to their enclosed design and propensity to generate prolonged zero-visibility conditions, backdrafts, and flashovers, enclosed spaces can become extremely dangerous during the course of an interior fire attack.

High-rise hallways

A high-rise hallway is essentially a horizontally enclosed space. It has also been documented that firefighters can become disoriented within the confines of a space as simple and familiar as a hallway when there is a complete and sustained loss of vision (fig. 3–18). As a result of an inability to promptly ventilate or evacuate without the guidance of a handline or vision from a thermal imaging camera, high-rise hallways should be considered extremely dangerous during a high-rise fire.

Fig. 3–18. In complete darkness, encountering loops of hose, handline separation, or self-locking hallway doors can hamper the ability of firefighters to reach fresh breathing air or the safety of a stairwell.

High-rise stairwells

A high-rise stairwell is essentially a vertically enclosed space in which firefighters, in zero-visibility conditions, can become disoriented with respect to which floor they are on. Therefore, the loss of direction occurs vertically in a stairwell as opposed to horizontally, as in the case of a high-rise hallway. For this reason, stairwells should be considered extremely dangerous during a fire.

While attempting to reach and ventilate a high-rise stairwell by opening a roof-access hatch, four Philadelphia firefighters became disoriented, ran out of air, and died. During an effort to obtain assistance from a rapid intervention team, the distressed company officer, blinded by very heavy smoke conditions, inadvertently gave command the incorrect floor number where they were located (Routley, Jennings, and Chubb 1991).

Large Enclosed Structure Defined

Enclosed structures can be built either over a basement or on a concrete slab foundation. They can be occupied, unoccupied, vacant, or abandoned during the course of a fire. If they are not protected by an automatic sprinkler system, they should be considered extremely dangerous when involved in a fire. Additionally, a life safety hazard may or may not exist during the incident. One of the most common examples of a large enclosed structure is commonly referred to as a "big box" structure (fig. 3–19).

Fig. 3–19. This large enclosed commercial structure ("big box" store) is fully protected by an automatic sprinkler system.

Although they may be built in several different configurations, these larger structures generally are defined as measuring approximately 100 ft. × 100 ft. or greater. Use of this specific measurement is beneficial as it serves two simple yet important purposes. First, it helps a firefighter to define the word "large" in quantifiable terms, and second, it serves as a vivid reminder that large enclosed structures measuring approximately 100 ft. × 100 ft. or greater have been responsible for numerous, multiple firefighter fatalities. One such incident took place in Goldsboro, North Carolina, in 1998. Unprotected large enclosed structure fires have historically been associated with multiple firefighter fatalities when a quick and inaccurate initial size-up was done, followed by a quick and aggressive interior attack. In other words, many firefighters did not survive these incidents when an initial, offensive strategy was used. The following is a summary of an incident that took place in the 1998 Goldsboro fire.

The Goldsboro incident took place in an enclosed commercial structure measuring 100 ft. × 100 ft. and not protected by an automatic sprinkler system. With light smoke showing on arrival, the owner informed the first-arriving officer that the auto salvage warehouse was unoccupied. After opening a sidewall for ventilation, in limited visibility conditions of 5 ft., two engine companies advanced handlines into the large structure in search of the seat of the fire. After locating and attacking the fire, firefighters were ordered to exit the structure, believing that the fire had been knocked down. As the nozzles were closed, firefighters were hit by a violent blast of heat that knocked them off their handline, immediately causing them to become disoriented. Transmissions for help were made, and one firefighter was rescued; however, additional rescue attempts were called off due to deteriorating conditions. Ultimately, two firefighters were unable to safely evacuate the building before running out of air. In both cases, the firefighters were located facing away from the point of entry with no indication of entrapment or entanglement by contents or debris (NIOSH 1999b).

Although an enclosed structure is described as having very few windows or doors to provide prompt ventilation and emergency evacuation, firefighters must remember a reality that contradicts the logic established in this definition. Regardless of the number of windows or doors (means of egress) present along the perimeter of a building, any large structure should be considered enclosed because it possesses the same degree of danger as an enclosed structure. The reasoning behind this has to do with numerous and sobering realities about large enclosed structure fires, including the risk associated with dangerous distances from the point of entry to the seat of the fire; the risk associated with exposure to life-threatening hazards during a deep-seated fire; that provision of safe, prompt, and effective ventilation may not be possible; and that following advancement of attack lines to the seat of the fire, egress or emergency evacuation of firefighters may not be accomplished promptly from a large enclosed structure. Therefore, when determining whether a structure is or is not an enclosed structure, firefighters must also consider the number of readily penetrable windows and doors of adequate number and size present *in relation to the length, width, and height of the structure.* In other words, the size of the structure must be considered not only in lateral terms but also vertically, and must be envisioned during a worst-case scenario involving a deep-seated working fire producing prolonged zero-visibility conditions. When firefighters did not consider or appreciate these dangers, it often resulted in firefighter injury or death.

Large Enclosed Structure Risk Classification Criteria

In the effort to prevent operations that overlook the true extent of danger during large enclosed structure fires, it is important for all firefighters to correctly and accurately assess risk associated with any large enclosed structure. First, as is the case involving small- and moderate-size enclosed structures, not all large enclosed structures are extremely dangerous. Some in fact are only "dangerous." When considering the risk classification of large enclosed structures during either prefire-planning activities or during an actual fire, firefighters must always remember that it is *the presence or absence of an operable sprinkler system* that ultimately determines the degree of danger associated with the structure and, therefore, the chances of becoming disoriented. In addition, firefighters must realize that they may not be able to differentiate a large enclosed structure that is dangerous from one that is extremely dangerous by sight alone. For example, considering that a structure is of new construction and should therefore have a sprinkler system may not always be an accurate method of assessing the structural risk associated with these structures. In general, when large enclosed structures are of newer construction in which modern building codes were followed, they are considered only dangerous. Depending on the occupancy, fuel type, load, and the type of construction, these codes typically require the installation of automatic sprinkler systems and, where needed, standpipe systems, fire department connections, smoke alarms, manual pull stations, and/or heat detectors. However, there may be other conditions involving the operability of the sprinkler system that may make a visual assessment inaccurate. For example, a fire may not be hot enough to activate a sprinkler system to control a fire and therefore fail to perform as anticipated, allowing heavy smoke to fill the structure. This would result in firefighters encountering an extremely dangerous fire in a large, unprotected enclosed structure. In another example, the wind may cause a break-up of the water pattern from activated sprinkler heads, rendering the heads ineffective in an extremely dangerous setting. A third example may involve an incident in which a sprinkler valve has been closed, leaving an entire fire floor of occupants and firefighters unprotected. A fourth example may involve a fire that is accidentally started by plumbers sweating pipe after the structure's sprinkler system was shut down. A fifth example may involve a structure that was originally equipped with a sprinkler system, but as expansion of the structure occurred or a change in occupancy type took place, the effectiveness of the original flow from the system was rendered inadequate. Finally, large enclosed structures that are either old or new can be fully protected by an automatic sprinkler system, but the piping may be clogged with sediment or calcification, preventing water from flowing when the heads activate. Therefore, large enclosed structures having operable sprinkler systems are classified as dangerous, meaning fatalities occur occasionally, while large enclosed structures without operable sprinkler systems are classified as extremely dangerous, meaning traumatic firefighter fatalities and, at times, multiple firefighter fatalities take place repeatedly during these high-risk structure fire operations. For familiarity purposes, the series of photographs in figures 3–20, 3–21, 3–22, 3–23, 3–24, 3–25, and 3–26 are but a few examples of large enclosed structures that are considered dangerous, with one exception, followed by some examples that are considered extremely dangerous when fast, aggressive, and uncalculated interior attacks are initiated.

Fig. 3–20. This is a new, protected, large enclosed structure housing an appliance store. The occupied structure has a risk classification of only dangerous.

Fig. 3–21. This is a new, large enclosed structure housing a department store. The risk classification at this point is unknown.

Preventing Firefighter Disorientation

Fig. 3–22. On closer inspection, fire department connections indicate that this structure is protected and considered only dangerous.

Fig. 3–23. This is a new, protected, large enclosed structure housing a commercial occupancy. The risk classification is dangerous, meaning occasionally fatal to firefighters.

Fig. 3–24. This is a new, protected, large enclosed structure housing a place of assembly. The risk classification is dangerous.

Fig. 3–25. This is a newer large enclosed structure housing a commercial occupancy. Since it has no sprinkler system, the risk classification for the building is "extremely dangerous," meaning it repeatedly takes the lives of firefighters.

Preventing Firefighter Disorientation

Fig. 3–26. This newer, large enclosed structure, which serves a detention center, is fully protected and considered only dangerous.

Keep in mind when visually assessing large enclosed structures that the basis for the extremely dangerous classification has to do with the absence of an automatic sprinkler system. Sprinkler systems must be operable in order for the structure to be protected by them and be considered only dangerous. The extremely dangerous structures are typically of older construction built prior to code adoption, but there may be exceptions. For example, newly constructed structures may not be protected by an automatic sprinkler system due to the occupancy type or the lack of code enforcement in unincorporated jurisdictions. Figures 3–27, 3–28, and 3–29 show some examples of extremely dangerous large enclosed structures.

Types of Enclosed Structures

Fig. 3–27. This is an older, vacant, two-story, unprotected, large enclosed structure. It is classified as extremely dangerous, meaning that firefighter fatalities occur in these types of structures at a disproportionate rate when an uncalculated offensive strategy is used.

Fig. 3–28. This is an older, unprotected, large enclosed structure occupied by a grocery store. The risk classification is extremely dangerous.

Fig. 3–29. This is an older, vacant, two-story, large enclosed structure with a basement that does not have a sprinkler system. It should be considered extremely dangerous and noted as such during prefire planning activities, enclosed structure tactics review, or a working fire.

References

Mora, W. R. 2003. "U.S. Firefighter Disorientation Study 1979–2001." http://www.sustainable-design.ie/fire/USA-San-Antonio_Firefighter-Disorientation-Study_July-2003.pdf.

National Institute for Occupational Safety and Health (NIOSH). 1999a. "Two Firefighters Die and Two Are Injured in Townhouse Fire, District of Columbia." http://www.cdc.gov/niosh/fire/reports/face9921.html.

———. 1999b. "Two Volunteer Firefighters Were Killed and One Firefighter and One Civilian Were Injured During an Interior Attack in an Auto Salvage Building—North Carolina." http://www.cdc.gov/niosh/fire/reports/face9832.html.

Routley, J. G. 1995. "Three Firefighters Die in Pittsburgh House Fire, Pittsburgh, Pennsylvania" (USFA-TR-078/February 1995). http://www.usfa.fema.gov/downloads/pdf/publications/tr-078.pdf.

Routley, J. G., Jennings, C., and Chubb, M. 1991. "Highrise Office Building Fire: One Meridian Plaza, Philadelphia, Pennsylvania" (USFA-TR-049/February 1991). http://www.usfa.fema.gov/downloads/pdf/publications/TR-049.pdf.

4

TYPES OF OBSCURED VISIBILITY CONDITIONS AND TYPES OF FIREFIGHTER DISORIENTATION

Zero Visibility Conditions

Smoke conditions encountered at structure fires are not all alike. In fact, significant differences in the quality and quantity of the smoke produced exist at every structure fire. With a greater awareness of these distinct differences, firefighter safety at structure fires can be enhanced.

It is widely understood in the firefighting community that smoke generated from a structure fire can be toxic, carcinogenic, flammable, and even explosive. Much has been written to properly inform firefighters of the multiple hazards of smoke produced from a structure fire. However, by gaining an understanding of the danger associated with the quality and the duration of smoke conditions, every firefighter will be more aware of one of the greatest hazards linked to firefighter disorientation. Of equal importance, these conditions also provide company and command officers with qualitative and quantitative bases to manage the risk.

Firefighters routinely operate in zero visibility conditions during the urgency of a working structure fire. A *zero visibility condition* is defined as blinding smoke conditions lasting less than 15 minutes in duration. Objectively speaking, zero visibility conditions should be considered dangerous, as they occasionally result in firefighter fatalities.

In the interior of an involved structure such as a single-family dwelling with an opened design, a firefighter wearing self-contained breathing apparatus (SCBA) may be able to easily see through light or moderate smoke conditions. On other occasions, firefighters will be able to see interior objects only on a very limited basis. The ability to see the interior arrangement of a structure, if only on a very limited basis, enables a firefighter to maintain a sense of direction, thereby allowing the firefighter to work effectively while still maintaining orientation. Poor or limited visibility conditions, although not significantly advantageous, may still allow the minimum required vision for the glow of the fire to be detected and for an attack on the fire to be successfully initiated.

Limited visibility conditions consequently necessitate the use of other senses to safely operate in the structure. Listening for the possible crackling sound of an active fire or the sound of shattering glass may help the firefighter find the seat of the fire. The sense of feel also may enable the firefighter to sense the heat penetrating through bunker gear, thereby enabling the firefighter to actually locate and extinguish the fire. Additionally, and in the absence of a thermal imaging camera, the sense of feel through gloved hands is a common technique used when conducting a primary search for incapacitated victims in limited visibility conditions. These interior firefighting conditions are well known, familiar to, and anticipated by the trained firefighter.

On the other hand, prolonged zero visibility conditions (PZVCs) are quite different, generally unfamiliar, at times unanticipated, and significantly more dangerous to the untrained firefighter. In fact, the modern firefighter should consider PZVCs to be one of the most dangerous hazards of interior structural firefighting because, as history has shown, it repeatedly takes the lives of multiple firefighters during single fire events.

Prolonged Zero Visibility Conditions (PZVCs)

Prolonged zero visibility conditions (PZVCs) are defined as blinding smoke conditions lasting 15 minutes or more (Mora 2003). Smoke conditions of this quality and duration should be considered extremely dangerous, as they have been repeatedly shown to result in the disorientation and fatality of firefighters. These conditions are not as common as zero visibility conditions nor are they as familiar to the untrained and inexperienced firefighter. During PZVCs, the single greatest hazard is the inability to see any object in any direction within a structure for the duration of the incident. In these conditions, the effective amount of breathing air contained in SCBA available to the firefighter is reduced to approximately 15 minutes in a 30-minute rated cylinder. The limited quantity of air combined with PZVCs produces a very dangerous environment should handline separation, entanglement, and disorientation occur. The 15-minute duration is based on an SCBA's effective breathing time given the higher oxygen consumption of firefighters during high-exertion activities. According to Routley, et al. (2008), "The rated duration of an SCBA is based on an average flow of 40 L/min [liters per minute]. Firefighters engaged in high-exertion activities are estimated to consume air at an average rate of 80 L/min and a maximum rate of 100 L/min. A 30-minute rated cylinder contains approximately 1,200 liters of usable air. This volume of air would be consumed in 15 minutes and a flow rate of 80 L/min."

A 1979 triple-disorientation incident occurring at a place of assembly in San Antonio, Texas, generated PZVCs lasting longer than seven hours. The incident ultimately required a defensive operation to control and extinguish, according to Albert Ersch, former captain of the San Antonio Fire Department and first engine officer to arrive on the scene. Furthermore, it is very important for all firefighters, especially the company, safety, sector, and command officers, to understand that the speed at which dangerous PZVCs are generated can vary significantly during the course of an incident. PZVCs can be found as companies arrive on the scene and therefore be obvious to firefighters. However, PZVCs can also occur very gradually, or at the other extreme, they can develop suddenly during an ongoing structure fire. PZVCs can therefore place interior firefighters in a highly hazardous environment that is primed for the possibility of disorientation.

Types of Firefighter Disorientation

Analysis of the disorientation study determined that there were in fact six different types of disorientation or six different ways in which firefighters lose visibility and their sense of direction during an interior structural attack. For safety, it is very important for all firefighters to understand that in all types of disorientation, the firefighter loses vision for the duration of the incident as a result of being exposed to a life-threatening hazard. *If firefighters can routinely avoid the life-threatening hazards common to all interior firefighting operations, they will avoid the disorientation that commonly leads to serious injury and firefighter fatality.* The following are the six types of firefighter disorientation:

1. Disorientation secondary to PZVCs
2. Disorientation secondary to flashover
3. Disorientation secondary to backdraft
4. Disorientation secondary to collapse
5. Disorientation secondary to wind-driven fire
6. Disorientation secondary to conversion steam

Disorientation secondary to prolonged zero visibility conditions (PZVCs)

Disorientation secondary to PZVCs is the most common form of disorientation and the type firefighters are most familiar with. During this event, the smoke conditions encountered on arrival may be light, moderate, or heavy. As the operation begins, during a quick primary search or a search for the seat of the fire, the light or moderate smoke conditions initially encountered, which may have provided some visibility, will deteriorate to limited visibility conditions. The resulting heavy and blinding smoke conditions caused by the burning contents or the burning structure itself, combined with a lack of effective ventilation, will continue to be produced and will prevent any visibility for the firefighters who have entered the structure originally or who subsequently enter the structure as backup or relief. If firefighters in this environment become separated from a handline or encounter entangled handlines, disorientation will occur. If assistance is not immediately provided, the firefighter or firefighters will not be able to safely reach a means of egress and fresh air. They will then deplete their air supply and may suffer serious injuries or die of asphyxiation and or thermal injuries.

"In 11 of 17 cases examined or 65%, firefighters exceeded their air supply while attempting to evacuate the structure" (Mora 2003). To reemphasize, one of the many dangerous and deceiving variables associated with the generation of smoke during enclosed structure fires is the speed with which PZVCs may develop. The blinding smoke conditions may be encountered on arrival or may develop rapidly during the incident.

In 1999, at the Worcester, Massachusetts, Cold Storage Warehouse fire, firefighters found light to moderate smoke conditions on arrival at a six-story, enclosed-structure fire involving a vacant cold storage warehouse. However, during the interior attack and over the span of only 2 minutes, visibility dropped from moderate to very heavy smoke conditions within the structure. The sustained blinding smoke was a contributing factor in the loss of six career firefighters.

Smoke may also completely fill a structure gradually over the course of an incident, as shown in figure 4–1. These variables must be known and anticipated by a trained incident commander observing the operation from a distance and by safety officers during the incident, either of whom may call for the evacuation of interior firefighters should the conditions within the structure warrant.

Following a cautiously determined decision to initially attack the seat of the fire, periodic progress reports on interior conditions provided by the interior sector or company officer and the effectiveness or ineffectiveness of the attack and safety of firefighters will dictate if a withdrawal from the structure is warranted.

Fig. 4–1. In 2007, Charleston firefighters found heavy smoke showing from the D-side of this large enclosed furniture store. Light smoke, however, was visible on the interior. During the interior attack, visibility gradually deteriorated as heavy blinding smoke, seen venting from the A-side, resulted in the deaths of nine firefighters, who were unable to find a hoseline to lead them to the point of entry. Firefighter disorientation caused by PZVCs was a contributing factor in the fatalities. (Courtesy Stewart English.)

Disorientation secondary to flashover

One of the leading causes of serious injury and fatality at both opened and enclosed structure fires is flashover. "A critical point in room fire growth is an event often referred to as flashover. While a universal definition does not exist, this event is generally associated with rapid transition in fire behavior from localized burning of fuel to involvement of all combustibles in the enclosure. Experimental work indicates that this transition can occur when upper room temperatures are between 400°C and 600°C (750°F and 1,112°F)" (Budnick and Evans 1986).

4 Types of Obscured Visibility Conditions and Types of Firefighter Disorientation

Unfortunately, this phenomenon consistently takes the lives of firefighters during fast and aggressive interior attacks. During an exposure to flashover, a firefighter will be exposed to a tremendous amount of heat from the engulfing fire. However, it is the blinding characteristic of the fire itself during a flashover that causes firefighter disorientation and the critical loss of direction.

There are definitive time intervals in which firefighters may be exposed to disorienting life-threatening hazards during which they must act to survive. By understanding these time frames, firefighters will in turn understand why certain types of disorientation and fatalities can occur not only in large but also in very small or medium-sized structures. For example, although the disorientation experienced following an exposure to PZVCs in a large warehouse generally plays out over a slightly longer period of time (typically the 15-minute breathing allotment firefighters have in their 30-minute-rated SCBAs), this is not the case in flashover or backdraft conditions. These sudden episodes require the firefighter to immediately exit or be effectively covered by protective streams during an emergency evacuation. Furthermore, if the firefighter is not able to exit the structure right away through a nearby door or window, the firefighter could die of thermal exposure during the struggle to exit. (Though it should be noted that firefighters have also survived exposures to flashover.)

The National Fire Protection Association (NFPA) has reported that a fully bunkered firefighter has approximately 17.5 seconds to leave a room after flashover before second-degree burns will be sustained. However, firefighters should not use this as a perceived margin of safety during interior operations, but rather as an understanding of the consequences of this type of exposure should they fail to exit within the 17.5-second time frame. For safety and survival, exposure to flashover should simply be avoided since personal protective equipment is not designed or intended to provide sustained thermal protection against this hazard. Disorientation secondary to flashovers occurs in both opened and enclosed structure fires, regardless of whether the structure is a well-sealed, commercial building or a single-family dwelling designed with adequate windows and doors for prompt evacuation (fig. 4–2).

Preventing Firefighter Disorientation

Fig. 4–2. In 2000 in Pensacola, Florida, with heavy smoke showing at an opened structure involving a residence, the first-arriving engine company advanced a 1¾-in. hoseline through the front door. As a flashover occurred, one of three evacuating firefighters became disoriented by the blinding fire and died after failing to exit. He was subsequently located at the interior C-D corner of the structure. (Courtesy of NIOSH 2000-44.)

The following incident happened in Cincinnati, Ohio:

> On March 21, 2003, a 25-year-old male career firefighter was fatally injured in a flashover during a house fire. The victim and two other fire fighters were on an interior attack crew and had just gone through the front door of a single-family residence. The hose line was uncharged and the crew was calling for water when a flashover occurred.... After the flashover, fire fighters on the front porch witnessed the victim walk toward the front door then turn and retreat into the structure. The two other fire fighters on the interior crew exited through the front door. They were injured and transported to the hospital where they were treated and released. The victim was located and removed from the structure within 10 minutes. He was transported via ambulance to the hospital where he was pronounced dead. (National Institute for Occupational Safety and Health [NIOSH] 2005a)

Other well-documented cases of disorientation caused by flashovers are on file and can be reviewed as an effective learning tool in the prevention of similar exposures.

Disorientation secondary to backdraft

A *backdraft* is defined as an "instantaneous explosion or rapid burning of superheated gases that occurs when oxygen is introduced into an oxygen-depleted confined space" (International Fire Service Training Association 2008, p. 122). During this condition, and when air is subsequently reintroduced, a very rapid fire in the form of a backdraft explosion results.

Although generally regarded in the fire service as a rare event, the effects of a backdraft can cause firefighters to suddenly become disoriented and injured. Backdrafts are typically violent and, like flashovers, will blind firefighters by fire or by a combination of smoke and fire. Like a flashover, the heat that is generated will also hamper firefighters. These exposures may cause firefighters to become separated from the handline that is serving as the critical lifeline out of the structure. Similar to PZVCs and flashovers, backdrafts and the disorientation they cause pose a serious life-threatening hazard that must be avoided at all costs. It is important for firefighters to clearly understand that, when conditions are right, a backdraft can occur in any confined space in both opened and enclosed structures. However, history has shown that when a backdraft takes place in an enclosed structure, it is usually fatal to firefighters. These types of fatalities have taken place in places such as Chicago, Boston, and Crooksville, Ohio.

Knowledge of the conditions that are ideal for a backdraft to take place will help firefighters to understand where and when the hazard can occur and, more importantly, what action should be taken to prevent or minimize exposure. As with all cases of disorientation, firefighters should be familiar with specific case studies involving disorientation secondary to backdraft. (Case studies of backdraft incidents are discussed in chapter 8.)

Disorientation secondary to collapse

A partial or complete collapse of a fire-weakened first floor in an opened structure with a basement or an enclosed structure with a basement will cause immediate disorientation. Generally, heavy fire and smoke will develop upward from the basement as air is introduced when the firefighter(s) drop through the burned opening in the floor. This occurs because the opening creates a vent point for the release of heat, smoke, and fire. In these scenarios, it is the fire or combination of fire and smoke that causes the loss of vision. In addition to becoming disoriented, the firefighter(s) who drop into involved basements are also exposed to the extreme heat of the fire and possibly entrapment caused by falling structural materials and contents.

A partial or complete collapse of a roof weakened by fire or the pressure of a backdraft in an opened structure, an opened structure with a basement, or an enclosed structure with a basement will also cause disorientation. Additionally, disorientation of firefighters can occur in rooms adjoining collapsed or collapsing areas as smoke, heat, and fire are pushed along pathways by the pressure of displaced air caused by the weight of the collapsing roof. During these cases, firefighters may be exposed to multiple life-threatening hazards. Not only will firefighters become disoriented by the fire, smoke, or the combination of fire and smoke, but they may also become entrapped by the weight of fallen ceiling materials, roofing materials, and possibly the load of heavy roof-mounted heating, ventilating, and air-conditioning equipment.

Simply put, in structures in which roofs or floors have collapsed, the interior firefighters may not survive unless they are quickly located, protected against the fire by high-flow handlines, and removed from the structure by an adequate number of personnel. If fatalities resulting from the type of scenarios described above are to be prevented, firefighters must avoid the life-threatening hazards of PZVCs, flashovers, backdrafts, and collapses.

One roof collapse incident took place in Houston, Texas, that caused the disorientation and burn injuries of five firefighters and the fatality of one officer (fig. 4–3). During the fire, Houston firefighters were exposed to a partial roof collapse after initiating a fast attack at a vacant, opened structure residence, 28 ft. wide by 64 ft. long. As the nozzle operator from E46 became disoriented by the heavy fire and smoke caused by the collapse, he moved toward the front of the house. At some point, he noticed flashing lights from the apparatus parked on the street and dove headfirst out a living room window on the D side. Other firefighters who became disoriented were assisted from the structure as well (NIOSH 2005b).

Fig. 4–3. This is a view of the front of 8510 Brandon at the time the roof collapsed, which pushed smoke and burning gases throughout the building. Crews from E46 and L46 were inside the structure at this time. (Courtesy of Houston Fire Department and Texas State Fire Marshal's Office Firefighter Fatality Investigation 05-218-02.)

Disorientation secondary to wind-driven fire

Research conducted on wind-driven fires in high-rise structures by Daniel Madrzykowski and Stephen Kerber of the National Institute of Standards and Technology has added to our knowledge base. Further research has shown that wind-driven fires can take place not only in high-rise structures but also in structures of any height, such as single- or two-story residences. Based on NIST research findings, a wind-driven fire is an extremely dangerous structure fire in which a

10-mph or stronger wind speed pressurizes fire venting from a window, door, or roof, causing the fire to rapidly spread when an interior flow path is established. NIST testing revealed that a vent point for a driven fire could be a high-rise stairwell doorway. In a scenario unfolding during the course of a high-rise fire, the stairwell doorway would actually be the point of entry for advancing firefighters onto the fire floor, but unfortunately, it could also serve as the dangerous vent point at the end of a flow path originating at a vented window in a fire-involved apartment (fig. 4–4). This can also describe the movement of fire during wind-driven fire conditions on grade level, regardless of whether or not the structure has an opened or an enclosed structural design.

Fig. 4–4. A stairwell doorway typically serves as the point of entry for advancing firefighters during high-rise incidents. However, during wind-driven fires, and when the door to the unit on fire is opened, it acts as a dangerous vent point for smoke, gases, and fire—at times having blowtorch characteristics.

Disorientation secondary to conversion steam

Although rare, all firefighters must be aware that injuries and fatalities of firefighters initiating aggressive interior attacks have occurred during enclosed structure fires, as a result of the disorientation that may be caused by the conversion steam produced during certain firefighting operations. Cases of disorientation secondary to conversion steam have taken place in large, nonresidential, enclosed structures such as carpet and paper warehouses where the trapped steam generated by activated sprinkler heads or by fire streams directed onto the fire produced heavy volumes of steam that blinded firefighters (fig. 4–5). Again, although rare, firefighters must be aware of the disorientation hazards during fires that have the potential to produce heavy quantities of conversion steam for prolonged periods of time.

Fig. 4–5. Even though a large enclosed commercial structure may be fully protected by an automatic sprinkler system, firefighters should nonetheless exercise caution due to the blinding and disorienting effects of conversion steam.

References

Budnick, E. K., and Evans, D. D., 1986. "Hand Calculations for Enclosure Fires." In *Fire Protection Handbook*, 16th ed., edited by A. E. Linville and J. L. Cote, 21–21. Quincy, MA: NFPA.

International Fire Service Training Association. 2008. *Essentials of Firefighting*, 5th ed. Stillwater, OK: Fire Protection Publications, Oklahoma State University.

Mora, W. R. 2003. "U.S. Firefighter Disorientation Study 1979–2001." http://www.sustainable-design.ie/fire/USA-San-Antonio_Firefighter-Disorientation-Study_July-2003.pdf.

National Institute for Occupational Safety and Health (NIOSH). 2005a, January. "Career Fire Fighter Dies and Two Career Fire Fighters Injured in a Flashover During a House Fire—Ohio" (F2003-12, Date Released: January 7, 2005). http://www.cdc.gov/niosh/fire/reports/face200312.html.

———. 2005b, December. "Career Fire Captain Dies When Trapped by Partial Roof Collapse in a Vacant House Fire—Texas" (2005-09, Date Released: December 16, 2005). http://www.cdc.gov/niosh/fire/reports/face200509.html.

Routley, J. G., Chiaramonte, M. D., Crawford, B. A., Piringer, P. A., Roche, K. M., and Sendelbach, T. E. 2008. "Firefighter Fatality Investigative Report, Sofa Super Store, 1807 Savannah Highway, Charleston, South Carolina." City of Charleston Post Incident Assessment and Review Team Phase II Report. http://downloads.pennnet.com/fe/misc/20080515charlestonreport.pdf.

5

SIMILARITIES IN DISORIENTATION FIRES

The only known similarity with respect to the specific structure fires examined during the course of the disorientation study, "U.S. Firefighter Disorientation Study 1979–2001," was that firefighters became confused or disoriented within the structure in each incident (Mora 2003). From that point, each incident was analyzed to determine if any other commonalities existed that would shed light on the causes or contributing factors that lead to disorientation.

It was initially observed that the disorientation fires did not seem to have anything in common. One unrelated characteristic involved the fact that they were of different types of occupancies. For example, the breakdown by occupancy of the 17 structures in which disorientation occurred consists of one place of assembly, one office building, one multiple-family dwelling, two single-family dwellings, three warehouses, three high-rise apartment buildings, and six commercial structures.

The construction types also varied and included five of fire-resistive construction, two of heavy timber construction, seven of unprotected noncombustible construction, and three of protected wood-frame construction.

The overall structural size also varied widely, ranging from a two-story townhouse unit (with a basement) that was 19 ft. wide by 33 ft. long, to a paper warehouse that was 500 ft. wide by 600 ft. long by 25 ft. high.

Regarding age, the structures were not built during any one particular era. The structures included a 20-year-old single-family dwelling; a thrift store that was originally a movie theater, constructed in the 1940s; and a 98-year-old, six-story warehouse built in 1905.

About half of the structures were occupied during the fire; half were unoccupied; and one, the Worcester, Massachusetts, Cold Storage Warehouse, although thought to be occupied, was vacant.

In summary, there were a wide range of different structural features associated with the 17 buildings examined, involving size, height, types of occupancy, and construction. Therefore, this finding, in and of itself, showed that disorientation can occur in structures of different sizes, heights, types of construction, and types of occupancy. Moreover, the structures in which firefighters became disoriented were occupied, unoccupied, or vacant during the course of a fire. However, additional analysis revealed that there were in fact nine major similarities that provided valuable insight into understanding the underlying contributing factors that lead to firefighter disorientation, and that will help prevent the disorientation of firefighters in the future. The following information touches on a few aspects of each similarity.

- **Structure: 100% occurred in an enclosed structure.** Prior to the study, the author had personal knowledge of four structure fires in which firefighters became disoriented. During the objective process of trying to determine similarities, it was strongly suspected that the enclosed structural design was common. This suspicion was confirmed, as disorientation incidents during the study time period (1999–2001) continued to take place nationally involving structures and spaces (high-rise hallways and basements) having an enclosed design. Following the collection and analysis of the data, 17 (100%) of the structures examined were in fact ultimately found to have an enclosed structural design.

- **Smoke: 94% had nothing, light, moderate, or heavy smoke showing on arrival.** In 94% of the cases, arriving firefighters encountered smoke conditions having various densities. These included some instances in which initially nothing was showing on arrival, but upon further investigation, smoke was encountered. In the remaining individual case (6%), fire was showing from the structure when the first fire company arrived on the scene.

- **Attack: 100% used an aggressive interior attack.** In all of the cases, firefighters initially executed an aggressive interior attack to locate and extinguish the seat of the fire. Various emergency tactics were also implemented during the incidents: In four cases (24%), a primary search was conducted. In two cases (12%), firefighters searched for fire extension. In three cases (18%), firefighters located and attacked the fire with handlines. In 12 cases (71%), the strategy changed from an offensive to a defensive operation, but only after injuries or fatalities had occurred.

- **PZVCs: 100% developed prolonged zero visibility conditions (PZVCs).** Although during the early stages of certain fires there may have been no smoke showing or light, moderate, or heavy smoke showing, nonetheless, in 100% of the cases PZVCs ultimately resulted. Some of these blinding conditions were initiated by the sudden onset of violent flashovers, backdrafts, and, in two cases, by rapid fire spread generated by wind-driven fires in two high-rise apartment buildings.

- **Handlines: 100% experienced handline separation or entangled handlines.** The circumstances surrounding the issue of handline separation and entangled handlines varied. According to the disorientation study, in 100% of the cases, firefighters became separated from the handline or became confused when they encountered loops of hose or tangled handlines. In one case, a single loop formed in the hose was enough to cause a firefighter to become disoriented and suffer serious thermal injuries in his attempt to exit an enclosed residential structure involving a converted convenience store. In this incident, the distance between the disoriented firefighter and the point of entry was 10 ft. In four cases (24%), firefighters did not advance a handline into the structure when conducting a primary search or when searching for the seat of the fire. In three cases (18%), firefighters were knocked off their handlines by the pressure created from flashovers and backdrafts. In one case (6%), a firefighter had the nozzle inadvertently pulled from his hands during an evacuation. In one case (6%), a firefighter became separated from the handline after falling through a fire weakened floor and into an involved basement. In one case (6%), a firefighter released the handline while evacuating, possibly to investigate a personal alert safety system device that was activating. In 41% of the cases, firefighters became confused when encountering tangled handlines in zero visibility as they attempted to evacuate the structure.

- **Company integrity: 100% experienced loss of company integrity.** In 100% of the cases, company integrity was lost in various ways. In some of the cases the loss of company integrity was intentional, while it was unintentional in others. In one case (6%), a fire officer was not with a crew or a buddy while inside the structure. In one case (6%), company integrity was lost as company members anxiously walked up a stairwell. In one case (6%), the crew did not enter the structure together. In one case (6%), company members became separated during the course of firefighting. In three cases (18%), company integrity was lost due to the force of flashovers or backdrafts. In 10 cases (59%), companies lost integrity during the evacuation from the building.

- **Disorientation: 100% experienced disorientation.** In 100% of the cases examined, firefighters experienced disorientation for sustained periods of time as they became blinded by the fire or smoke from exposure to flashovers, backdrafts, wind-driven fires, or collapses of floor or roofs, or by the development of heavy conversion steam. In 11 cases (65%), firefighters depleted their air supply while attempting to evacuate. In four cases (24%), firefighters were trapped due to collapsing floors or collapsing truss roofs. In three cases (18%), company integrity was lost due to the force of flashovers, backdrafts, or wind-driven fire.

- Disorientation also occurred within the structures at various distances measured from the point of entry. Disorientation occurred in distances of 10 ft., 10–20 ft., 25 ft., and 30–40 ft. from the point of entry. In two overly aggressive attacks, advances were made over distances of as much as 80 and 200 ft., resulting in disorientation.

- **Sprinklers: 88% lacked a functioning sprinkler system.** In 15 structures (88%), there was no sprinkler system, or the system failed to operate. In one case (6%), a sprinkler valve was closed, leaving a high-rise hallway unprotected. In one case (6%), the heat of the fire was not great enough to activate the sprinkler heads during a paper warehouse fire, and the structure's ventilation system did not operate. In two cases (12%), the system did activate. During one of these operations, however, high winds that entered a high-rise hallway rendered the activating sprinkler heads ineffective. Another incident occurred in a large mill warehouse that involved the firefighting and rescue efforts of Lieutenant Clifford Clement of the Fall River (MA) Fire Department. In addition to PZVCs generated during the fire, Clement advised that water spray from activating sprinkler heads mixed with suspended powder from a dye-processing operation, creating a film of "blinding soup" that covered all personal protective equipment including hand lights and the masks of self-contained breathing apparatus.

- **Disorientation sequence: 100% involved a firefighter disorientation sequence.** In each of the 17 incidents examined during the study period, a chain or sequence of events was identified that ultimately caused firefighters to become disoriented. Furthermore, 27 additional disorientation sequences taking place in enclosed structure fires across the country were also identified following a review of cases included in the U.S. Fire Administration's "U.S. Firefighter Fatality Retrospective Study" (TriData Corporation 2002).

References

Mora, W. R. 2003. "U.S. Firefighter Disorientation Study 1979–2001." http://www.sustainable-design.ie/fire/USA-San-Antonio_Firefighter-Disorientation-Study_July-2003.pdf.

TriData Corporation. 2002. "U.S. Firefighter Fatality Retrospective Study, April 2002/FA-220." (Prepared by TriData Corporation for Federal Emergency Management Agency, United States Fire Administration, National Fire Data Center.) http://www.usfa.fema.gov/downloads/pdf/publications/fa-220.pdf.

6

THE FIREFIGHTER DISORIENTATION SEQUENCE

In the aftermath of a serious firefighter injury or fatality, a chain of events leading to the unfavorable outcome can be identified. One specific chain revealed during the course of the disorientation study and associated with 100% of the cases analyzed is referred to as the firefighter disorientation sequence (Mora 2003).

Simply defined, the firefighter disorientation sequence is *a chain of events leading to firefighter disorientation.* A key step in the sequence that eliminates a firefighter's ability to see within the structure involves the development of and exposure to a life-threatening hazard. These are the same hazards that ultimately cause firefighters to become disoriented and injured or killed.

Just as there are different types of firefighter disorientation, there are also different types of firefighter disorientation sequences that develop at varying speeds and cause sustained loss of firefighter visibility. The different types of firefighter disorientation sequences or chains are associated with the specific hazards that help to generate them. These life-threatening hazards, familiar to firefighters, are often described on the fireground as "deteriorating conditions" or "rapidly deteriorating conditions." Based on reported exposures to life-threatening hazards during disorientation fires, the various firefighter disorientation sequences and the general speeds in which the involved hazards develop are listed in the following table.

Preventing Firefighter Disorientation

Type of Firefighter Disorientation Sequence	Speed of Hazard Development
Zero visibility sequence	Slow, gradual, or rapid, involving prolonged zero visibility conditions (PZVCs)
Flashover sequence	Sudden, involving flashover
Backdraft sequence	Sudden, involving backdraft
Collapse sequence	Sudden, involving collapse
Wind-driven fire sequence	Sudden, involving wind-driven fire
Conversion steam sequence	Slow to gradual, involving conversion steam

Regarding the disorientation sequence involving prolonged zero visibility, incident analysis has shown that the speed of PZVC development is related to the amount and type of fuel load, size of the structure, and whether air is introduced into the structure during the incident. PZVCs can develop rapidly in a small structure or, as in the case of a larger structure, slowly, gradually, or rapidly.

For survival, it is crucial for firefighters to be aware of the steps that form the disorientation chain. The chain of events involves firefighters who (mistakenly) quickly and aggressively execute offensive strategies and tactics when conditions encountered incorrectly indicate they are safe to implement. In general, the steps in the firefighter disorientation sequence are the following (Mora 2003):

1. Fire in an enclosed structure with smoke showing
2. A fast and aggressive interior attack
3. Deteriorating visibility conditions
4. Handline separation or entangled handlines encountered
5. Disorientation

The Out-of-Sequence Chain

Although the steps of the disorientation sequence are rather simple and generally coincide with standard operating procedures used across the country when initiating an offensive strategy, the firefighter disorientation sequence can unfold without regard to the usual order of events. In other words, the disorientation sequence can play out in a different order during an incident. For example, instead of a logical sequence of events from 1 to 5, firefighter disorientation can also fatally unfold in a 1,2,4,3,5 series. In other words, step 4, handline separation, can happen before conditions within a structure deteriorate, as occurred at a large enclosed structure fire involving a paper warehouse in 1999 in Kansas City, Missouri. This possibility of an out-of-sequence chain underscores the importance for every interior firefighter, regardless of assignment, to remain in physical contact with a handline, which can serve as a lifeline out of an enclosed structure fire should PZVCs eventually set in during the course of an operation.

During the Kansas City incident, a battalion chief, who was first to arrive on the scene of a reported fire at the paper warehouse, investigated to find a light haze of shifting smoke in the structure. He determined that multiple handlines could control a fire that was located in a large pile of paper in the center of the massive structure. However, interior conditions gradually deteriorated during the unsuccessful attempt to extinguish the fire, and 52 minutes after arrival, blinding smoke completely filled the structure. The battalion chief, who was acting as the interior sector officer, was last seen donning his facemask. However, he was working alone, deep within the structure, and without the benefit of a handline to serve as a lifeline out of the structure. Unfortunately, the officer was not located before he ran out of air, even though he was in constant radio contact with the command post and several rapid intervention teams (RITs) were repeatedly pressed into service (National Institute for Occupational Safety and Health [NIOSH] 1999).

Firefighter disorientation may occur in typical sequence

1. Fire in an enclosed structure with smoke showing
2. A fast and aggressive interior attack
3. Deteriorating conditions
4. Handline separation or entangled handlines encountered
5. Disorientation

Firefighter disorientation may occur out of sequence

1. Fire in an enclosed structure with smoke showing
2. A fast and aggressive interior attack
4. Handline separation
3. Deteriorating conditions
5. Disorientation

Firefighter disorientation involving the out-of-sequence handline separation step also tragically played out in Charleston, South Carolina, in 2007. Although only light smoke was initially visible at the right rear portion of the store's interior, conditions of heavy, blinding smoke developed. Catastrophically, only those firefighters who were capable of maintaining contact with a handline or who were guided out by other firefighters in contact with a handline were able to exit and survive (NIOSH 2009).

Out-of-sequence disorientation also occurred during the Worcester, Massachusetts, Cold Storage Warehouse fire of 1999. During the course of the incident, six firefighters who were not in contact with a handline were exposed to PZVCs. The thick smoke that filled the structure disoriented the firefighters, who then ran out of air while attempting to relocate their points of entry (NIOSH 2000).

These complex, enclosed structure fires underscore the importance of understanding that opened structure fires (which are small to moderate in size and have a means to promptly ventilate and evacuate) are evaluated and controlled differently during an interior fire attack than enclosed structure fires. Simply stated, unlike opened structure fires, enclosed structure fires cannot in all cases be safely and universally managed using offensive strategy and tactics involving fast, aggressive, and blind interior attacks from the unburned side. It is widely understood, for example, that following an accurate initial size-up, during working opened structure fires, an engine company can attack the seat of the fire while other firefighters from an arriving truck company simultaneously conduct a primary search or search for extension without a handline and do so without exposure to unreasonable risk.

On the other hand, fires developing in enclosed structures first require that the risk be carefully evaluated and weighed and that tactics and tasks be implemented differently. Otherwise, as overwhelming documented evidence has clearly shown, it may be fatal to firefighters.

While an aggressive interior attack was underway on the fifth floor at the C-D corner of a Houston high-rise building in 2001, a cold front with winds of 16 knots gusting to 39 knots entered the city. As two officers retraced handlines while attempting to exit the floor, they were exposed to PZVCs, entangled handlines, and extreme heat as the fire was driven into the hallway by the wind entering the involved room through opened windows. Company integrity was lost as one officer, disoriented by the blinding smoke, ran out of air and died before being rescued by an RIT. In the process, two members of the RIT also became disoriented during the rescue attempt at this extremely dangerous enclosed structure incident, as shown in figure 6–1 (Texas State Fire Marshal's Office 2001).

It is important to note, therefore, that the generally accepted practice that does not require firefighters to maintain close proximity to or continuous contact with a handline during a smaller, opened structure fire involving, for example, a single-family residence, cannot be applied when an enclosed structure fire is involved, whether small, moderate, or large in size.

Fig. 6–1. This is a view of the open windows on the C-D corner of the building, which allowed high winds to force fire into the involved unit and hallway. (Courtesy of Houston Fire Department.)

To prevent the pivotal handline separation step of the disorientation sequence from taking place, every firefighter must be made aware of the problem and be required to maintain close proximity to or constant contact with a handline when entering, working on the interior, or exiting fire conditions involving an enclosed structure of any given size.

Conversely, firefighters must also understand that if a link in the chain can be broken, the disorientation that leads to an injury or fatality during an enclosed structure fire can be prevented. Although the handline separation step is only one of the critical steps to avoid, for firefighter safety, it would be good practice to attempt to avoid all links in the firefighter disorientation sequence, thereby making it difficult for the sequence to ever completely unfold. That starts by recognizing the different types of enclosed structures during the prefire planning or initial size-up process.

References

Mora, W. R. 2003. "U.S. Firefighter Disorientation Study 1979–2001." http://www.sustainable-design.ie/fire/USA-San-Antonio_Firefighter-Disorientation-Study_July-2003.pdf.

National Institute for Occupational Safety and Health (NIOSH). 1999. "Warehouse Fire Claims the Life of a Battalion Chief—Missouri." http://www.cdc.gov/niosh/fire/reports/face9948.html.

———. 2000. "Six Career Fire Fighters Killed in Cold-Storage and Warehouse Building Fire—Massachusetts." http://www.cdc.gov/niosh/fire/reports/face9947.html.

———. 2009. "Nine Career Fire Fighters Die in Rapid Fire Progression at Commercial Furniture Showroom—South Carolina." http://www.cdc.gov/niosh/fire/reports/face200718.html.

Texas State Fire Marshal's Office. 2001. "Texas State Fire Marshal's Office Line of Duty Death Investigation Number 02-50-10, Captain Jay Jahnke, Houston Fire Department, October 13, 2001." http://www.tdi.texas.gov/reports/fire/documents/fmloddjahnke.pdf.

7
INITIAL SIZE-UP FACTORS

An initial size-up is an important, rapid evaluation of fireground factors that serves as the basis for decisions made by the first-arriving officer at the scene of a structure fire. Although not all-inclusive, the following list contains some of these factors:

- Time of day
- Weather
- Type of structure (opened or enclosed)
- Type of occupancy
- Type of construction
- Age of the structure
- Condition of the structure
- Amount of smoke showing on arrival
- Amount of fire showing on arrival
- Amount of time the fire has been burning

Misreading Initial Size-Up Factors

Heavy emphasis is customarily placed on the accuracy of an initial size-up and of conducting an ongoing size-up to keep abreast of changing fireground conditions. During the early stages of an incident, an initial size-up is quickly conducted. The size-up is a rapidly applied and highly logical procedure; the success of an operation typically hinges on the speed and accuracy with which the first-arriving officer conducts the size-up. For example, if a size-up is not conducted rapidly and a decision not formulated to quickly and aggressively attack the fire offensively, the outcome may be unfavorable. Some results of a slow size-up process and subsequently slow decision making to initiate a fast and aggressive interior attack may include the following:

- Occupants may be fatally injured if firefighters do not enter quickly and remove them from the structure.
- A flashover may occur, engulfing firefighters and occupants.
- A fire-weakened roof may collapse, trapping advancing firefighters and occupants.
- Property loss sustained by fire, smoke, and water damage may be excessive.

On the other hand, if the initial size-up reveals that it is too risky or impossible for fully protected firefighters to enter the involved structure due to the volume of fire and the unsafe nature of structural conditions, then the decision will be made to conduct a defensive attack. In this strategy, the attack is made from the exterior while maintaining safe collapse zones and protecting any threatened exposures.

Generally speaking, the preceding brief scenarios accurately describe the first-arriving officer's thought process during the critical first seconds of a structure fire. However, analysis conducted during the disorientation study revealed a serious inconsistency (Mora 2003). According to the study:

> Established Standard Operating Procedures in United States Fire Departments call for firefighters to utilize a quick and aggressive interior attack; also referred to as an Offensive Strategy, at any structure fire safe to enter. One fireground condition included in an officers' initial size up, is the amount of smoke showing from the structure on arrival. Light, moderate, or heavy smoke are common smoke conditions in which an officer would initiate a quick interior attack, to locate and extinguish the seat of

> the fire or to conduct a primary search. Based on study results, however, these were the same smoke conditions that were found on arrival at 94% of the enclosed structure fires that ultimately resulted in disorientation.

Further, according to the study, when traditional initial size-up factors were seen by arriving firefighters and translated into fast and aggressive interior attacks into structures and spaces (basements) having an enclosed design, in 100% of the cases, disorientation, serious injuries, narrow escapes, and the deaths of 23 firefighters resulted. In other words, the initial size-up factors observed at the scene of the enclosed structure fires examined had an entirely different meaning and were therefore seriously and unknowingly misinterpreted.

In comparison, whenever light, moderate, or heavy, but tenable smoke is seen venting from an opened structure fire, it usually indicates that conditions are safe enough to initiate an offensive strategy to search for and attack the seat of the fire or to perform a primary search. However, when the same factors were observed at the scene of the enclosed structure fire in the study, they indicated the opposite—that it was not safe to implement an aggressive interior attack. Therefore, in the setting of an enclosed structure fire, these initial size-up factors in fact mean that there is a possibility that advancing firefighters may become exposed to life-threatening hazards either in the form of a flashover, a backdraft, a collapse of the roof or floor, or an exposure to prolonged zero visibility conditions, any one or a combination of which could cause firefighter disorientation leading to fatalities.

Recall that in 100% of the cases examined in the disorientation study, firefighters made a fast and aggressive interior attack, an indication that the firefighters were not aware of the extreme danger associated with the enclosed structural design or of the actual meaning of the familiar initial size-up factors observed. In this regard, every firefighter involved in every case examined was operating within a set of widely accepted strategic and tactical assumptions held by virtually every department in the nation. The problem continues to plague the fire service to this day. In addition to understanding that common size-up factors as seen on arrival from either an opened or enclosed structure have a corresponding meaning, it is also beneficial to understand which specific type of structure fire takes the lives of firefighters most often, as well as understanding the safety, effectiveness, and reliability of the tactics utilized at those fires.

Analysis of the Structural Firefighter Fatality Database

The disorientation study (Mora 2003) attempted to learn what caused firefighters to become disoriented and what could be done to prevent it. In the process, it was revealed that structures that had either an opened or enclosed design were linked to life-threatening hazards that caused firefighter disorientation, although it appeared that disorientation occurred in opened structures only occasionally. However, and in quantitative terms, it was generally unknown which of the two structural designs was more dangerous during a fire.

During the course of interior attacks, traumatic fatalities can occur during any structure fire. However, the fatalities that occurred while using a fast and aggressive interior attack have been observed to occur more frequently in enclosed structures and spaces. In fact, it has been conclusively determined that structures and spaces that possess an enclosed structural design are significantly more dangerous to the safety of firefighters. This is because enclosed structures are more prone to produce the life-threatening hazards that are known to contribute to sustained disorientation and the inability to promptly ventilate and evacuate the structure, which then leads to fatalities. This was confirmed in findings of an analysis of structural firefighter fatalities conducted on data spanning a 16-year time span (January 1, 1990 to December 31, 2006). Data for the analysis were provided by the U.S. Fire Administration. (U.S. Fire Administration National Fire Data Center and National Fallen Firefighter Foundation 2000). Note: The color-coded Analytical Database used can be accessed by contacting the United States Fire Administration at www.usfa.dhs.gov. The following sections are the contents of the structural firefighter fatality analysis.

Background

In an effort to more accurately determine the degree of risk associated with opened and enclosed structure fires, an analysis was conducted of 444 cases of firefighter fatalities occurring while on the scene of structure fires. Examination of the extensive information found in the database provided by the U.S. Fire Administration National Fire Data Center proved helpful in determining valuable information pertaining to the number of firefighter fatalities that occurred in opened and enclosed structure fires and of the tactics utilized. This key database may also serve as an excellent source of information to address other causes of line-of-duty deaths and as a baseline for comparison in determining progress over time in the reduction of line-of-duty deaths in each of the fatality categories.

Information sought

At a minimum, the information sought in this analysis included the total number of line-of-duty deaths and their distribution between the opened and enclosed structures fires that occurred during the 16-year time span. To ensure an accurate count of the specific types of traumatic injuries sought, structure fires listed were color-coded to differentiate the types of injuries that led to fatalities during each incident.

Criteria

The criteria used during each incident analysis included the following:

    a. The focus was on those structure fires in which an interior attack or primary search was initiated and which ultimately resulted in traumatic firefighter fatalities.

    b. The injuries that caused the line-of-duty deaths were required to be a result of the types of life-threatening hazards firefighters may encounter in the interior of a structure fire, including flashover, backdraft, a partial or complete collapse of the roof, collapse of a floor into a basement, or prolonged zero visibility conditions.

After careful study of the details in each summary, those that met the criteria were categorized by color.

Color coding for structure type, action, and outcome

Red—Enclosed structure fires with interior attacks and traumatic fatalities.

Yellow—Opened structure fires with interior attacks and traumatic fatalities.

Pink—Type of structure could not be determined with certainty. However, an opened or enclosed structural design would have had nothing to do with the fatalities under the criteria. These included fatalities that occurred as a result of a chimney falling onto firefighters during overhaul, an exterior wall falling outward onto firefighters during exposure protection, a canopy or façade falling onto firefighters, and firefighters crushed by apparatus.

Turquoise—Heart attacks or cerebrovascular accidents (CVAs).

Blue—Electrocutions.

Findings

The findings of the "Analysis of Structural Firefighter Fatality Database" (Mora 2007) are as follows: Of the 444 firefighter fatalities taking place "while on the scene of structure fires," the criteria were met in 123 structure fires. The 123 structure fires resulted in 176 traumatic firefighter fatalities during an aggressive interior attack.

Of 123 total structure fires:

89 (72%) involved enclosed structure fires.

34 (28%) involved opened structure fires.

Of 176 total firefighter fatalities:

135 (77%) firefighter fatalities occurred in enclosed structure fires.

41 (23%) firefighter fatalities occurred in opened structure fires.

Of 38 total multiple firefighter fatality structure fires:

32 (84%) occurred in enclosed structure fires.

6 (16%) occurred in opened structure fires.

Conclusion

Over a 16-year time span, firefighters utilizing an interior attack died far more often, in greater numbers, and with greater multiple line-of-duty deaths in enclosed structure fires than in opened structure fires. All structure fires are to be considered dangerous and may be fatal to firefighters, although enclosed structure fires are significantly more dangerous than opened structure fires (Mora 2007).

Although it was determined that when traditional fast and aggressive interior attacks were utilized, enclosed structure fire operations were more dangerous than those conducted in structures with opened designs. This strongly suggests that at times either defensive strategies or tactics that are implemented differently must be used to avoid exposure to life-threatening hazards both during opened and enclosed structure fires. In simpler and broader terms, the fire service has an enormous enclosed structure problem, but it also has an ongoing opened structure problem to contend with. Should disorientation occur within an opened or enclosed structure fire, firefighters must understand that there is a high probability that it will lead to smoke inhalation, burns, or trauma, or a combination of these injuries, and result in fatalities. Therefore, the only reasonable operational approach for firefighters to take is to use tactics that will routinely avoid the hazards that are known to lead to firefighter disorientation.

References

Mora, W. R. 2003. "U.S. Firefighter Disorientation Study 1979–2001." http://www.sustainable-design.ie/fire/USA-San-Antonio_Firefighter-Disorientation-Study_July-2003.pdf.

———. 2007. "Analysis of Structural Firefighter Fatality Database." http://www.iafc.org/associations/4685/files/safeWkResFFsurvField09_AnalysisOfStr-FFFDatabase.pdf.

U.S. Fire Administration National Fire Data Center and National Fallen Firefighter Foundation. 2000. Analytical Firefighter Fatality Database. Accessed by contacting the U.S. Fire Administration, www.usfa.dhs.gov.

8

AVOIDING LIFE-THREATENING HAZARDS

Backdraft

A backdraft is "an instantaneous explosion or rapid burning of superheated gases that occurs when oxygen is introduced into an oxygen-depleted confined space" (IFSTA 2008). A violent and life-threatening hazard that can cause serious injuries to firefighters working within a structure, a backdraft can also cause firefighter disorientation. Therefore, firefighters must understand the conditions that lead to this rare yet potentially fatal event.

According to the National Fire Protection Association's *NFPA 555: Guide on Methods for Evaluating Potential for Room Flashover*, "Fires occurring in normal scenarios, in which air is available, will transition into flashover which may directly engulf firefighters as they advance into a structure in search of the seat of the fire or victims." *NFPA 555* continues to explain that during the course of interior fires in tightly closed spaces, the oxygen concentration will drop from the normal 21%. When the concentration of oxygen is reduced to 8%–12% and if there is sufficient heat remaining in the space, it will represent an ideal condition for a violent backdraft to occur should additional air be intentionally or unintentionally introduced into the space (NFPA

2013a). Additionally, it must be understood that under certain conditions violent backdrafts may occur when air is reintroduced into any oxygen-depleted space such as a basement, ceiling space, attic space, stairwell, stairway, and, on a larger scale, an entire enclosed structure.

In 2004 in Pittsburgh, Pennsylvania, an electrical fire originating in the ceiling space of a basement beneath a large enclosed place of worship resulted in a violent backdraft (National Institute for Occupational Safety and Health [NIOSH] 2006). The force of the backdraft ejected firefighters out of the building, causing multiple injuries. After knockdown and during extinguishment of hot spots, the structure's steeple collapsed and fell and trapped two firefighters in rubble and debris, causing their fatalities. Although there was no disorientation experienced, the unprotected large enclosed structure nevertheless took the lives of two firefighters and injured 29 other firefighters following an interior attack (fig. 8–1).

Fig. 8–1. An electrical fire originating in the ceiling space of a basement below this large enclosed place of worship resulted in a violent backdraft. The force of the backdraft ejected firefighters out of the building, causing multiple injuries. (Courtesy of NIOSH F2004-17 [see NIOSH 2006].)

Actions to avoid backdraft

Trained and experienced firefighters are well aware of the potential backdraft hazard and understand that they are in fact rare events. However, firefighters must realize that although backdrafts can occur in any structure, opened or enclosed, an enclosed structure or space provides the airtight environment ideal for producing disorienting and life-threatening backdrafts. Therefore, due to this structural condition, firefighters should be prepared for a backdraft especially in these structures.

Achieving safety during backdraft conditions requires a concerted effort by every involved firefighter to identify the signs of an imminent backdraft and to take correct action to avoid exposure. As in any firefighting operation, a total team effort to maintain safety on the fireground is absolutely necessary, including the recognition of signs and immediate exchange of information regarding dangerous conditions observed. Many incidents involving exposures to backdrafts occurred when life safety was not an issue. Due to established procedures, firefighters may have been exposed to excessive risk during particularly dangerous operations.

Life-threatening hazards including backdrafts may be avoided by remembering and routinely following recommendations offered by such organizations as the International Association of Fire Chiefs (IAFC) and National Fallen Firefighters Foundation (NFFF). The IAFC's Rules of Engagement for Structural Firefighting advise: "Do not risk your life for lives or property that cannot be saved" (IAFC 2012). Initiative 4 of the NFFF's 16 Firefighter Life Safety Initiatives stipulates, "All firefighters must be empowered to stop unsafe practices" (NFFF 2011). As an example of Initiative 4, given the urgency of a working structure fire, it is common for firefighters to still need to complete the process of bunkering up as they arrive on the scene. On occasion during this brief span of time, the driver may have an uninterrupted moment to conduct an initial size-up and detect the signs of a backdraft. However, if the probability of injury or firefighter fatality is to be effectively reduced, the signs of a backdraft must be recognized by someone—anyone—at the onset or during the course of an incident and quickly communicated to the officer in charge so that firefighters may take safe and corrective action.

Exterior and interior signs of backdraft

Exterior signs that typically indicate conditions associated with a backdraft have been well documented and can be used to predict a backdraft explosion.

In their report, "The Current Knowledge & Training Regarding Backdraft, Flashover, and Other Rapid Fire Progression Phenomena," Gregory E. Gorbett and Ronald Hopkins provide firefighters with clear indicators of a backdraft.

The following are indicators that a backdraft may occur.

1. The fire may be pulsating. Windows and doors are closed, but smoke is seeping out around them under pressure and being drawn back into the building.
2. No visible flames in the room.
3. Hot doors and windows.
4. Whistling sounds around doors and windows. If the fire had been burning for a long time in a concealed space, a lot of unburned gases may have accumulated.
5. Window glass is discolored and may be cracked from heat.

The key indicator that has been witnessed in the past is the in and out movement of the smoke, which gives the appearance that the "building is breathing" (Gorbett and Hopkins 2007).

These are signs seen from the exterior that indicate that a backdraft is about to occur. During these conditions, firefighters must back away from the structure and prepare for an explosion that may dangerously propel debris, including broken glass, metal, and bricks, outward from the structure. If considered safe and if firefighters are properly equipped, a decision may be made to conduct vertical ventilation to vent smoke from the structure. However, keep in mind that air may be entrained inward through opened windows or doors at grade level during this ventilation process, initiating a smoke explosion or backdraft that may cause the roof—and firefighters—to be lifted upward by the force of the explosion. A review of the following case summary, which occurred in 2008 in Colorado, describes this potential hazard (NIOSH 2009b).

8 Avoiding Life-Threatening Hazards

The Colorado explosion incident

On February 22, 2008, a deputy chief and eight firefighters were injured during an explosion at a restaurant fire in Colorado. At 1340 hours, dispatch reported visible smoke and flames through the roof of a commercial structure. At 1344 hours, police arrived and began evacuating the restaurant and the adjoining retail store. The restaurant was part of a block-long row of adjoining structures. In 25 minutes, three engines, two ladder trucks, and 24 fire department members, including the injured firefighters, were on scene.

> A crew entered the restaurant with moderate smoke showing toward the rear and no flames visible. The crew backed out and entered the retail store (an adjacent building attached to the restaurant) to check for fire in the ceiling but found only light smoke visible. Another crew attempted to ventilate the retail store with a chainsaw, and when the roof was noticed to be spongy, they moved to the roof of the next building, two buildings down from the restaurant. Interior crews operating in all three buildings had backed out. A crew closed the front doors of the restaurant fearing the oxygen would feed the increasingly greenish-black smoke pushing out of the roof of the restaurant. Fire ground personnel noticed the front windows of the restaurant and adjoining retail store were vibrating as flames from the roof of the restaurant intensified. At 1427 hours, the restaurant and two adjoining buildings exploded, sending glass, bricks, and wood debris into the street. The crew on the roof located two buildings down from the restaurant, felt the front portion of the flat roof heave up about 5 feet, sending a fire officer to the ground below and temporarily trapping four other fire fighters; all incurred injuries. In addition, four fire fighters, positioned on the ground within 6 feet of the store fronts, were injured by flying debris.
>
> Key contributing factors identified in this investigation included fire growth and smoke buildup in the common

117

attic area of the buildings that pressurized and exploded, unrecognized building characteristics that contributed to the fire and explosion hazards, ineffective ventilation, execution of offensive operation SOPs [standard operating procedures] and inadequate staffing. (NIOSH 2009b)

During the course of certain past enclosed structure fires, exterior signs of a backdraft did not exist and therefore did not allow the prediction of a backdraft. On the other hand, interior signs were observed by surviving firefighters just before a backdraft event. This critical information, referenced from the NIOSH Fire Fighter Fatality Investigation and Prevention Program, can therefore be used in future operations to serve as interior and exterior backdraft warning signs and of the need to take precautionary action to avoid dangerous exposure (fig. 8–2).

The Crooksville, Ohio, backdraft incident

In 1998, five firefighters in Crooksville, Ohio, searching for the seat of the fire in the basement of an unoccupied opened structure, first observed "light to moderate gray puffing smoke and then lazy orange flames from the ceiling area" (NIOSH 1998b). The structure was a single-story family dwelling measuring 28 ft. by 50 ft. After quickly extinguishing the visible flames and in an attempt to check for fire spread in the ceiling space, a firefighter, using an axe, lifted a ceiling tile, initiating a violent and disorienting backdraft. Two of five firefighters were unable to exit through the ensuing blinding smoke and fire, even though two nearby means of egress were available. The "lazy flames" observed by the firefighters may have been signs of a reduced oxygen concentration within the enclosed ceiling space, as opposed to more active flaming commonly seen in rooms having higher oxygen concentrations.

The interior signs of imminent backdraft seen during the Crooksville incident included "puffing smoke" and "lazy flames" (NIOSH 1998b). Another obvious sign included the presence of an enclosed ceiling space within an enclosed basement, both of which are associated with fatal backdrafts.

Fig. 8–2. Best observed from a distance, one exterior sign of a potential backdraft explosion includes the venting of heavy pressurized smoke from every existing crevice or opening in a structure. (Courtesy of Tim Olk.)

A tactic that may be considered when interior backdraft signs of "lazy flames" and/or "puffing smoke" are seen along suspended ceilings or from walls on the interior of a basement involves completely evacuating the structure, venting, and, when possible, attacking the fire from the floor above the basement from a safe position, such as a doorway leading to the exterior. This would require fully bunkered firefighters and multiple handlines to attack the fire while others provide protective water cover for the attacking firefighters.

The objective of this tactic, completed without introducing additional air into the ceiling space, is to sufficiently vent the smoke and heat and to reduce the temperature in the ceiling space in order to eliminate the heat required to trigger a backdraft in the oxygen-depleted space. After a reasonable application rate and time, with additional charged handlines on standby, shut down the flow from the handlines and start to pull the ceiling just inside the point of entry to see if the fire in the ceiling space has been knocked down. Be cautious because of the possibility that lightweight trusses or beams supporting the first floor may have been weakened by exposure to the fire. If an assessment with a thermal imaging camera determines that this is not the case, then continue to pull ceiling tiles as advancement is made to extinguish any remaining hot spots. This tactic offers an alternative approach to the traditional fast and aggressive interior attack, where pulling ceiling tile from deep interior positions is linked to firefighter fatalities. In the process, potential backdrafts and weakening of the floor-ceiling assembly, which may cause a partial or complete collapse of the first floor, may also be prevented. This approach may also avoid the danger of becoming disoriented by the smoke and fire associated with a backdraft or a partial or complete collapse that could trap firefighters in the structure.

The Memphis backdraft incident

A large enclosed commercial structure fire that included a backdraft occurred in Memphis, Tennessee, in 1999. During this incident firefighters attacked heavy fire in an office located at the rear of the structure, while other firefighters popped suspended ceiling tiles to check for fire extension. However, like the Crooksville incident, a violent, disorienting backdraft occurred.

Aside from the presence of a suspended ceiling, no other interior signs that may have predicted a backdraft were reported (NIOSH 2004). Therefore, whether signs are seen or unseen, firefighters must understand that a ceiling space is an enclosed space that may cause oxygen concentrations to drop within the space during a fire. Reduced oxygen concentration may subsequently trigger a backdraft when air and therefore oxygen is intentionally or unintentionally reintroduced. The interior sign that a possible backdraft could occur during the Memphis incident included the presence of an enclosed suspended ceiling space within a large enclosed structure.

Tactically, and in an effort to avoid the disorienting effects of a backdraft deep within a structure, when an enclosed ceiling space is encountered, suspended ceiling tiles should be removed, but done so in close proximity to a point of entry. Thereafter and when needed, a ground ladder must be raised to a position such that interior access is not hampered and the entire ceiling space can be examined by scanning with a thermal imaging camera.

Signs of smoke or fire and the presence of lightweight trusses or unprotected steel beams that may elongate when exposed to fire and push outward on exterior walls must be noted and communicated to command. These conditions can confirm a working fire and the possibility of a partial or complete collapse of the roof.

If, as in the case of the Crooksville basement backdraft incident, fire is seen in the ceiling space from the point of entry, it could be attacked laterally above the ceiling space from the safety of a point of entry. This approach for checking a ceiling space at the point of entry, whether in smaller or larger structures, provides greater safety as it places firefighters adjacent to the exterior and to backup companies equipped with charged backup lines who could immediately provide rescue should a backdraft actually occur.

The Boston backdraft incident

Significant interior signs that preceded a backdraft were reported during a Boston structure fire in 2007. The incident involved an enclosed structure having an enclosed ceiling space and a basement. As Boston firefighters arrived at the scene of a vacated enclosed structure housing a restaurant, fire was seen venting through the roof. Several firefighters advanced handlines into the structure and extinguished fire that was visible above a stove in the rear kitchen area. During extinguishment, several *suspended ceiling tiles were dislodged.* During the next few minutes, and as a heavy layer of smoke banked downward within the structure, windows next to the point of entry were vented. These events were followed by a violent backdraft, which instantaneously disoriented and engulfed 12 firefighters who were operating in the interior. Two firefighters were unable to exit the structure (NIOSH 2009c).

The interior and exterior signs indicating that a possible backdraft could occur during the Boston incident included the presence of an enclosed ceiling space situated in an enclosed structure with a basement. Since they were enclosed, all areas of the structure—specifically, the first floor, the basement, and the ceiling space—provided the right environment for a backdraft to develop.

During this incident, another sign indicating the risk of a backdraft was that fire was seen, upon arrival, venting through the roof of the enclosed structure. In contrast, when fire is seen venting through the roof of an *opened* structure, it typically indicates that fire, heat, and smoke are therefore also venting from the structure through the burn hole and thus gives firefighters the opportunity to safely enter and make an effective attack on the seat of the fire from the interior. But because this was an enclosed structure, the presence of fire venting through the roof was a danger sign of potential backdraft.

In addition, with opened structures, even if a fire involves the ceiling space adjacent to the burn hole and a backdraft takes place, firefighters will typically be able to promptly evacuate the structure by means of available windows or doors. However, because of the lack of windows or doors, this is not the case with enclosed structures, especially large enclosed structures.

During this incident, an additional sign of a risk of backdraft was the heavy banking or mushrooming of smoke that filled the interior prior to horizontal ventilation by means of windows or doors. In this particular incident, and when mushrooming or banking of smoke is encountered at the scene of similar fires, certain steps should be taken. Due to the possibility of a backdraft, the structure's doors and windows should initially be kept shut to prevent air entrainment, which could precipitate a backdraft explosion. Streams from multiple handlines or an elevated master stream are then directed at the venting fire and into the burn hole in the roof to allow the water to extinguish the venting fire and to knock down any fire extending in the ceiling space. In the process, potential backdrafts may be avoided and potential weakening of the roof assembly that may cause a partial or complete collapse of the roof may also be prevented. This approach may also avoid exposure to prolonged zero visibility conditions (PZVCs), disorientation, fire, and deep and dangerous distances commonly encountered in larger enclosed structures, where firefighters would, because of a total and sustained loss of vision, be forced to crawl through debris and entanglements to reach points of entry and safely exit the structure.

With the Boston scenario in mind, it would be safest to use an enclosed structure standard operating guideline (SOG) where the risk is safely managed and executed within the limitations of the resources available. The details of enclosed structure tactics and operating guidelines for enclosed structure fires are described in chapter 12.

The Chicago backdraft incident

A backdraft incident that occurred in Chicago in 1998 involved an unoccupied auto repair shop. Nothing was showing as firefighters arrived at the scene of the enclosed structure. Several firefighters advanced a charged handline into the rear of the enclosed repair shop, which was tightly filled with numerous automobiles. These firefighters observed a heavy layer of black smoke that had banked down 3 ft. from the ceiling, indicating the presence of a fire. After the roof was vented, fire and heavy smoke appeared from the vent hole. During this time, interior firefighters searched for the seat of the fire as heavy smoke continued to bank down until it reached the floor. With zero visibility and after the roof had been vented, a nearby overhead door auto-activated, initiating a violent backdraft that immediately disoriented 10 of the firefighters who were in the room at the time (NIOSH 1998a).

The interior and exterior signs indicating that a possible backdraft could occur in the Chicago incident included the presence of an enclosed structure with a basement. Additional signs included fire found upon venting the roof and heavy smoke that banked down to the floor prior to ventilation, which was unintentionally caused by the raising of an overhead door.

As in the case of the Boston restaurant fire, when a large enclosed structure is involved with fire venting through the roof, a tactic to consider involves directly attacking the fire where it is seen. This should be the case whether fire is showing along the sides of or from the roof of the structure.

Steps previously discussed for managing conditions encountered during the Boston incident are also executed for fires similar to the Chicago fire. Due to the possibility of a backdraft, the structure's doors and windows should initially be kept shut to prevent air entrainment, which could precipitate a backdraft explosion. Streams from multiple handlines or an elevated master stream is then directed at the venting fire and into the burn hole in the roof to allow the water to extinguish the venting fire. In the process, potential backdrafts may be avoided, and potential weakening of the roof assembly, which may cause a partial or complete collapse of the roof, may also be prevented.

With the Crooksville, Memphis, and Chicago backdraft scenarios in mind, whenever the location of the seat of the fire within an enclosed structure is unknown, it would be safest to assume the worst-case scenario and to implement an enclosed structure SOG programmed to avoid the multiple life-threatening hazards enclosed structures are known to present. The details of the enclosed structure SOG utilized when the location of the seat of the fire is unknown are described in chapter 12.

Flashover

Flashover is defined as "a stage in the development of a contained fire in which all exposed surfaces reach ignition temperatures more or less simultaneously and fire spreads rapidly throughout the space" (NFPA 2013a). Unfortunately, firefighters are injured and killed by exposure to flashovers while conducting interior firefighting operations.

Furthermore, when disorientation cases were closely examined, it was seen that firefighters not only suffered external and internal burn injuries by exposure to fire, but they also suddenly became disoriented by the fire's blinding effects.

Because of the loss of vision caused by fire, firefighters are typically unable to see the door they used during entry and therefore often accidentally miss the closest means of egress to safety. During a sudden and urgent emergency evacuation, firefighters may run into other disoriented firefighters, collide with walls, and stumble over contents, slowing their escape from the structure. Furthermore, because of the critical loss of vision, evacuation time is dangerously extended when they are forced to crawl through the fire.

Nationally, circumstances surrounding fatal flashover scenarios are similar. More importantly, the repetition of flashover fatalities is a strong indicator that firefighters are not being appropriately trained to recognize when a flashover will occur and how to react to prevent exposure during an interior attack (fig. 8–3).

Fig. 8–3. Flashover is one of the most dangerous hazards in the fire service. When signs of imminent flashover are not recognized and appropriate action not taken, the result may include exposure to fire, blinding firefighter disorientation, and firefighter fatalities. (Courtesy of Dennis Walus.)

Avoiding the flashover hazard

To prevent exposure to flashover and subsequent disorientation, every firefighter must first be able to quickly recognize and know how to react to the signs of imminent flashover. During a structure fire, one warning sign of flashover is referred to as rollover, also known as flameover. As inferred, this involves the ignition of flammable fire gases collected along the ceiling of a room, often appearing as flames rolling over the surface of the ceiling. Should this sign be visible in a room, firefighters without the protection of a charged handline should not enter. However, should rollover appear while firefighters with a handline are in a room, the firefighters should immediately get low and direct a stream to the ceiling space to cool the room and prevent a flashover from taking place.

One associated problem that firefighters should be aware of is that while advancing, the flames signaling an imminent flashover may not be visible through heavy smoke from floor level. This was the condition found during a number of tragic fires where thick, blinding smoke prevented firefighters from seeing the fire overhead before conditions rapidly deteriorated. Flashovers that expose firefighters are devastating, as they frequently and instantaneously create multiple problems for firefighters in the interior of the structure. As a flashover occurs, engulfing the entire room from ceiling to the floor, fire may burn the company's hoseline in two, causing the loss of water pressure and flow. This event leaves the crew without the protection of a fire stream and the lifeline needed to reverse their direction of travel and safely crawl back to the point of entry. Additionally, firefighters will be instantaneously exposed to and suffer injury from the extreme heat of the engulfing fire, which will burn through personal protective equipment; blister, melt, and deform facemasks to further obscure vision; and damage other self-contained breathing apparatus (SCBA) components. However, in spite of this possibility, firefighters may still survive an exposure to flashover if they exit promptly.

A firefighter protected by an NFPA-compliant structural firefighting ensemble that includes a helmet, hood, coat, pants, coveralls, boots, and gloves has 17.5 seconds following flashover exposure before sustaining second-degree burns (NFPA 2013b). During this fire exposure, a firefighter may very well sustain second-degree burns; however, the firefighter may survive the event should a window or door allow an evacuation to take place. This scenario has played out on numerous occasions on the fireground. However, if there is no attempt to promptly evacuate or if disorientation prevents immediate exit, fatality will probably result. For this reason, it is good policy to routinely have charged backup lines advanced into the structure as soon as possible to knock down visible fire that may cut off points of entry; to attack hidden fire in ceiling spaces that may drop, cutting off the first-arriving company's means of egress; and to provide a protective cover for the first-arriving engine and/or truck company in the event of flashover.

The problem with the loss of vision due to the blinding effects of fire is that the time to react, and thus to survive, is brief, usually only a matter of seconds. This is unlike disorientation secondary to prolonged zero visibility, in which a firefighter has the time remaining in the air tank to either reach the safety of the exterior or to allow a rapid intervention team or teams to remove the disoriented firefighter from the structure.

The reality associated with disorientation caused either by flashover or backdraft is that these life-threatening hazards make an emergency evacuation difficult even from a small opened structure and almost impossible from an enclosed structure of any size, because of the enclosed architectural design and absence of accessible and readily penetrable windows or doors. Enclosed structures should be viewed in essence as death traps for overly aggressive and untrained firefighters.

Heavy, tenable smoke, although dangerous, is a condition that commonly fills a room or structure during firefighting operations and is safely and routinely managed during the course of a fire. However, whenever the smoke concentration in a room has deteriorated to zero or near zero visibility conditions and the temperature increases to the point of pain for a fully bunkered firefighter, these conditions signal that a flashover is about to take place. Although recognized to be very subjective measurements, the "hot and heavy" heat and smoke combination condition serves as a real-time warning to all involved firefighters that an impending life-threatening and disorienting flashover is imminent.

Firefighters must also remember that although present, fire at the ceiling level in many cases will not be visible to the naked eye, due to the blinding smoke. The onset of flashover may also be instantaneous and violent, leaving firefighters no time to decide on what to do. For this reason, tactical company contingencies are critical to survival and must be understood, used, and periodically reviewed by all crew members.

Flashover contingency action plans

There is no structure that is more valuable than the life of a firefighter. Therefore, for firefighter safety, a flashover contingency action plan for painfully hot temperatures associated with heavy smoke conditions that precede flashover and lead to disorientation should be understood and utilized. The following is one example of such a plan.

A flashover contingency plan. The nozzle operator must clearly understand that a flashover is imminent whenever the combination of heavy smoke and hot temperature, which has become painfully hot, causes bunkered firefighters to drop to the floor. Therefore in this situation, even without the ability to actually see the fire, the operator must fully open the nozzle and direct the stream to the ceiling, where flashover first begins. During this chain of events, firefighters will have no visibility within the structure and therefore will be relying on their sense of feel. As the heat of the fire sinks through bunker gear, the crew must understand that they are in a dangerous, imminent flashover situation and that the nozzle operator must, without hesitation, fully open the nozzle to prevent a disorienting and fatal flashover from engulfing them. In addition, and to prevent a premature shutdown of the nozzle, the flow from the handline must be continuous until the company officer orders the line to be shut down (fig. 8–4).

Although this flashover contingency action plan may cause some water damage, it will increase firefighter safety. Safety will be enhanced by preventing flashover, preventing the disorientation of firefighters, preventing the burning of the handline (which would result both in a loss of water and of a life line), and ultimately preventing a defensive operation possibly involving serious injuries or multiple firefighter fatalities.

8 | Avoiding Life-Threatening Hazards

Fig. 8–4. After conducting a 360° walk-around and considering all other size-up factors, firefighters don their gear to make entry into the structure. Since the light to moderate smoke showing from this single-family residence is not found to be untenable, an offensive attack will be initiated to locate and extinguish the seat of the fire. Because adequate staffing is on hand, a primary search, with the assistance of a thermal imaging camera, will also be simultaneously conducted. Whenever the temperature is painfully hot and the smoke condition heavy, firefighters are to use their flashover contingency plan, which calls for immediately opening the nozzle to prevent an impending flashover from engulfing them in the structure. (Courtesy of Dennis Walus.)

Balanced concern for water damage

Currently accepted tactics, although well intentioned, may in fact work against firefighter safety. Historically speaking, firefighters have been commonly given basic training on the correct procedures to follow in preventing or minimizing water damage during the course of a working structure fire. During these operations, firefighters have been instructed not to open the nozzle if the glow of the fire is not visible, as it would result in unnecessary water damage—a simple and understandable rule. For this reason, the usual rule was that the nozzle should only be opened to apply water when fire is actually visible. However, in the interest of preventing disorientation secondary to a blinding flashover, whenever an imminent flashover situation materializes, typically in zero visibility, the person on the nozzle should have the officers' standing approval, regardless of what firefighters are doing at the time, to point the nozzle toward the ceiling and fully open the nozzle (fig. 8–5). This emergency maneuver is intended to cool the room to prevent an engulfing flashover event from occurring. There should be no delay in this action. Subsequent closing of the nozzle should then be done when conditions stabilize and when approved by the company officer.

Fig. 8–5. Flashovers can occur at any time, even while interior firefighters may be focusing on other urgent tasks. For this reason, everyone must be continuously aware of the dangerous combination of hot temperatures in conjunction with heavy smoke conditions. (Courtesy of Dennis Walus.)

Since the specific combination of hot temperatures and heavy smoke is a sign of imminent flashover, firefighters must learn to anticipate and use this sign to prevent future fatal exposures and, in so doing, join the safety culture advocated by many organizations, which stresses that firefighters should never unnecessarily risk their lives to save a structure or the contents of a structure. In short, a concern to protect property must be balanced with a reasonable concern for firefighter life safety.

Consider the action that occurs in a room protected by an automatic sprinkler system from a safety perspective. During a fire, heads will activate and control the fire, prevent flashover, and allow arriving firefighters an opportunity to complete extinguishment under safer conditions. Therefore, there is a much lower chance of sustaining serious injury in a room or structure protected by an automatic sprinkler system. Similarly, in rooms that are in imminent flashover conditions and that are not protected by an automatic sprinkler system, fully opening a nozzle or nozzles to prevent a devastating exposure to a flashover is absolutely warranted and should be a standard approach in preventing the line-of-duty deaths attributable to the thermal and disorientation hazards resulting from flashover.

If firefighters are trained in the survival tactic that calls for automatically and fully opening nozzles whenever a flashover is imminent, injuries may be prevented and lives may be saved.

In the following section, and as an introduction to additional discussion on managing life-threatening hazards, the topic of fire dynamics including the important concepts of flow path and ventilation-induced flashover are discussed.

Fire dynamics

Throughout the years, firefighters have learned about extreme fire behavior to assess the dangerous conditions present or that may develop during the course of working structure fires. According to Daniel Madrzykowski, safety research engineer for the UL Firefighter Safety Research Institute (FSRI), "Fire dynamics can provide a firefighter with means to understand how a fire will grow and spread within a structure and how best to control that growth" (Madrzykowski 2013).

As a part of this learning process, Madrzykowski points out that today's methods of construction produce homes that are more tightly built and therefore result in a much different interior fire environment.

> As new homes retain heat and the gaseous fuels burn better than old homes and as synthetic fuels burn faster than wood and cotton, the probability of arriving to a pre-heated, fuel-rich fire environment has increased in recent

years. As a result, fires are controlled by the amount of oxygen available to them.

Under these conditions and in basic terms, control of the fire is initially accomplished on the fireground by preventing firefighters from opening doors or windows until a crew can advance a charged attack line to locate and attack the fire. Thereafter, the structure can be safely ventilated without unwanted fire development. In addition, and as a warning, Madrzykowski stresses,

> If a door or a window is opened while the fire is still burning, although at a reduced level, and if additional fuel is available, the introduction of outside air can result in a rapid increase in the heat-release rate of the fire and may enable enough energy generation to flashover the room. This transition has been referred to as a ventilation-induced flashover.

And although firefighters have not referred to this transition as such, they have witnessed ventilation-induced flashovers, many violent in nature, during firefighting operations in the past. In the future, this condition must be anticipated—underscoring that ventilation and entry must be delayed, as stated, until charged handlines or portable monitors are positioned for a coordinated attack on the fire, which will develop rapidly. If this important and well-timed tactic is overlooked, the quickly developing and spreading fires will cause firefighters to play "catch up" with the fire.

Flow path

The flow path is a very important aspect of fire dynamics. As described by Madrzykowski,

> The flow path is the volume between an inlet and an exhaust that allows the movement of heat and smoke from a higher-pressure area within the fire area towards lower-pressure areas accessible via doors, windows, and

other openings. Depending on its configuration, a structure can have several flow paths. Operations conducted in the flow path, between where the fire is and where the fire wants to go, places firefighters at significant risk due to the increased flow of fire, heat, and smoke toward their positions.

Should wind-driven fire conditions cause rapid fire spread, firefighters working in the flow path could be in extreme danger. In essence, when a flow path exists through the interior of a structure, the pressure created by the fire within a room will take the path of least resistance and vent out any available window or door. The same is true even with natural-ventilation cases, whether wind is present or not.

More information on flow path and fire dynamics research is available in Madrzykowski's *International Fire Service Journal of Leadership and Management* article "Fire Dynamics: The Science of Fire Fighting" (2013).

Ventilation-induced flashover

Firefighters are generally aware of the common flashover hazard, and fortunately most try to work in a manner that avoids it. But firefighters must also understand how a ventilation-induced flashover develops.

For example, a hazardous ventilation-induced flashover might occur in a fire-involved room located toward the rear of a structure that has not vented. Firefighters then initiate an aggressive interior attack from the front or unburned side of the structure. As the crew advances inward, the involved room ventilates, either following the breaking of windows caused by the heat of the fire or intentionally by horizontal ventilation efforts. As the room is ventilated, it may result in room flashover, which can cause rapid fire spread that can travel at velocities of 10 mph engulfing and overrunning the advancing fire crew (Madrzykowski and Kerber 2009).

Unlike the "hot and heavy" sign of the typical flashover, which does provide some degree of warning on the interior, the signs indicating rapid ventilation-induced flashover spread may not be seen, and therefore there may be no warning. Once the flow path has been created and the room has flashed over, firefighters in the exhaust portion of the flow path will be exposed to a very fast-moving wall of flames and/or hot fire gases. Therefore, the fact that a room is observed venting fire on arrival is in and of itself an exterior sign of the possibility that rapid fire spread toward the point of entry may take place. This may occur if an unrestricted flow path exists and an aggressive interior attack is initiated without first attacking the venting fire from the exterior for pre-entry cooling purposes, as suggested by NIST and Underwriters Laboratories (UL) studies.

Softening the target

Beginning in 2010 and continuing through 2013, NIST and UL conducted studies of single- and two-story residential structure fires and tactics. NIST and UL have tested structures where the fire was ventilation limited and then only one opening was made prior to suppression. They have also examined cases where the fire had spread to many areas of the structure, due to multiple exterior openings, and then the fire was attacked. Figures 8–6 to 8–11 are photographs documenting some of these cases during testing from 2010 to 2013.

Fig. 8–6. FDNY Firefighters take part in testing at a two-story structure with NIST and UL officials. The studies took place in 2012 on Governors Island, New York. (Courtesy of NIST.)

Fig. 8–7. Firefighters attack with a stream from a smooth bore nozzle. (Courtesy of NIST.)

Fig. 8–8. Heavy fire was showing from the front door of a single-story structure as Chicago firefighters participated in 2010 testing with NIST officials. (Courtesy of NIST.)

Preventing Firefighter Disorientation

Fig. 8–9. As the fire is attacked through the front door, various readings are recorded by NIST. (Courtesy of NIST.)

Fig. 8–10. Spartanburg firefighters observe heavy fire showing from a window during fire experiments with NIST and the International Society of Fire Service Instructors (ISFSI) officials. (Courtesy of NIST.)

Fig. 8–11. Spartanburg firefighters direct a straight stream at a steep angle off the ceiling to quickly cool the room and allow conversion steam to readily vent from the window opening. This method of water stream application attempts to enable conversion steam to effectively vent from the window and to the exterior instead of forcing the bulk of the steam into the interior as would have resulted had a semifog or full fog pattern been used. (Courtesy of NIST.)

Based on their findings, NIST and UL both suggest that, for safety, room fires that have not extended to involve the structure be initially attacked from the exterior using a straight stream pattern from an adjustable or smooth bore nozzle when fire is seen venting on arrival. For a single-room and contents fire with hot fuel gases filling the fire floor, a water flow of 100 gallons per minute (gpm) or more showed that an initial exterior attack utilizing a straight stream pattern directed on the ceiling for 15 seconds would significantly reduce the interior temperatures, to prevent flashovers prior to initiating an interior fire attack. It should also be noted that experiments were conducted with water flows ranging from 100 gpm to 180 gpm, with the majority of the experiments conducted by members of the FDNY flowing 180 gpm with a solid stream pattern via smooth bore nozzles. Stephen Kerber (2013) describes the need to "soften the target" in reference to a room and contents fire in a residential structure in his article, "What Research Tells Us About the Modern Fireground." When a fire in a single-story residential structure of moderate size with fire showing from windows and the door on the A side was considered, Kerber stated:

> Applying water to the fire as quickly as possible—regardless of where it is emitting from—can make conditions in the entire structure better. During the UL experiments, water was applied for approximately 15 seconds into a door or window with fire coming from it or with access to the fire from the exterior. This small amount of water had a positive impact on conditions within the structure, increasing the potential for victim survivability and firefighter safety.

The test results seen for the single-story residential structure, which indicated safer and faster fire knockdown when evaluated against the effort and time required to implement an offensive strategy, were similar for experiments conducted in a simulated two-story residential structure also involving a room and contents fire. The following scenario is described with a comparison to traditional tactics:

> Fire is showing from the second floor of side A. Traditional tactics call for the hoseline to be charged in the front of the house prior to entry, but water is usually not flowed onto the fire prior to entry. Even if the interior path to the fire is known, flowing water directly onto the fire is faster from the outside than it is from the inside. A common reason why this is not done is because the conditions beyond the fire would be made worse. In this experiment, temperatures were measured in the hallway just outside the room and in the other bedrooms on the second floor. Twenty-five gallons of water directed off of the ceiling of the fire room from the exterior decreased fire room temperatures from 1,792°F to 632°F in 10 seconds; the hallway temperature decreased from 273°F to 104°F in 10 seconds.

In light of these significant findings, the NIST-UL method of attack during "fire showing in room and contents" scenarios should be seriously considered as another alternative in the effort to avoid exposure to flashover and therefore to help prevent the disorientation, injuries, and fatalities that flashover is known to cause.

Self-Venting and Wind-Driven Fires

To gain a greater understanding of dangerous self-venting and wind-driven fires, Madrzykowski, conducted eight laboratory experiments that were performed in a simulated apartment at NIST. In email correspondence, Madrzykowski (2009) expressed how hazardous the thermal conditions can become during scenarios with no external wind (self-venting) or during 10–20 mph external wind speeds (wind-driven), as observed during the testing phase. According to Madrzykowski:

> It does not take a significant wind to create the "Blow Torch" results in the doorway in the flow path. In test 1, no external wind was imposed on the apartment, but "self-venting" of the window and a flow path through the apartment into a corridor and out a door-sized vent, resulted in 10 mph hot gas flow (floor to ceiling) coming out of the apartment doorway. This resulted in temperatures in the "flow path" in the corridor in excess of 1100°F from floor to ceiling. The "nonflow" side of the corridor had temperatures near 600°F at the ceiling and 400°F near the floor. Still too hot. And this is a case with no external wind applied. Bottom line: A 10–20 mph wind outside the structure causes a significant thermal hazard inside, but there are also cases where the fire gases and ventilation path within the structure can make enough flow velocity to create hazardous thermal conditions even for firefighters in full personal protective equipment.

It is therefore important for firefighters to be aware of both wind-driven fire as well as the self-vented room flashover hazard.

Tactical options available to firefighters during self-venting scenarios include a delay of horizontal ventilation, which would delay the venting of the room on fire. Another tactic may involve the previously discussed NIST-UL recommendation to initially soften the target by attacking the fire from the exterior as needed to provide interior cooling prior to entry. A final option used by interior firefighters who may suddenly encounter the hazard is to use the flashover contingency plan by fully opening the nozzle on the full fog setting to protect the crew from the rapidly spreading fire. For safety, a cautious retreat of the entire crew while still discharging water for protection during this event would be appropriate. Additionally, hose evolutions that would routinely provide a quick backup line would also be highly beneficial for firefighter safety during this and other unforeseen life-threatening hazards.

Wind control devices (WCDs)

To prevent exposure to the effects of wind-driven fires, whether in high-rise buildings or in single- or multiple-family residences on grade level, wind control devices (WCDs) such as fire-resistive fire curtains or fire blankets offer another alternative (figs. 8–12 and 8–13).

Fig. 8–12. Fire curtains measure 8 ft. × 6 ft. and are resistant to temperatures up to 2,000°F. (Courtesy of NIST.)

8 Avoiding Life-Threatening Hazards

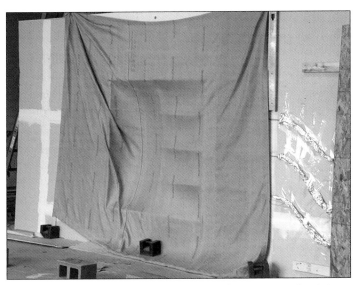

Fig. 8–13. Fire blankets measure 10 ft. × 12 ft. (Courtesy of NIST.)

According to NIST, "Wind Control Devices such as Fire Blankets or Fire Curtains reduce oxygen available to the fire and slows down the spread of flame, heat and toxic gases" (Kerber and Madrzykowski 2009b).

When used on the exterior during a pressurized fire venting from the C side of a two-story home, for example, fire curtains or fire blankets suspended from ground ladders may be raised to seal windows from pressure of the wind at the inlets, thereby allowing firefighters the chance to safely enter structures from the A side to conduct searches and attacks on the fire. However, the use of a flexible approach may provide the best outcomes and sound officer judgment must always be used. For example, in certain cases too many windows may be venting on the pressurized side of a residence on ground level, and the necessary number of curtains and ground ladders may not be readily available to quickly, safely, and effectively seal all the windows. The rapid use of water from handlines when available would then be the officers' first choice of action. It must be stressed that *operating above the fire to deploy WCDs should only be conducted in fire-resistive construction where floors of reinforced concrete separate operating firefighters from the fire immediately below—never in ordinary construction.*

In other cases, it may be best to use a combination attack with water and fire curtains. Fires at single- and two-story residences, which may produce high radiant heat from the venting of an entire side of a structure, may preclude the use of WCDs because of the risk of injury to firefighters. In these scenarios, and while no ventilation is allowed to take place on any side of the structure, including the front door, the use of multiple preconnected attack lines to quickly knock down the main body of fire on the venting side would be the first-arriving officer's initial course of action.

An exterior attack on fire that is venting from an opening on the pressurized side of a structure should be considered during both rescue and nonrescue scenarios because, as determined by the NIST, if an interior flow path has been established, advancing through an opening such as a door on the opposite side of the structure will create a vent point that will place firefighters and occupants in the dangerous flow path of rapidly spreading fire.

WCDs that are resistive to temperatures up to 2,000°F can be used by firefighters to seal structural openings, thus preventing wind from driving fire, heat, and smoke through the interior on any floor of any fire-resistive structure of any height.

Avoiding wind-driven fires

The blinding effects and thermal hazard associated with exposure to flashover also can occur instantaneously as a result of the impact of wind on a structure fire. These unique types of structure fires may also produce a hazardous volume of flames and hot fire gases moving through a flow path, from the high-pressure side to the low-pressure vents, at times having "blow torch" characteristics.

Findings associated with dangers of high-rise wind-driven fires have been determined and disseminated by NIST in a training DVD released in 2009 titled *Evaluating Fire Fighting Tactics Under Wind Driven Conditions*, by Steven Kerber and Daniel Madrzykowski (2009b). Additionally, other noteworthy wind-driven fire documents include, "Analysis of a Fatal Wind-Driven Fire in a Single-Story House" (Barowy and Madrzykowski 2012a) and "Wind Driven Fire Research: Hazards and Tactics" (Madrzykowski and Kerber 2010).

Because of the serious safety hazard linked to these conditions, these findings must be translated into safe and effective action on the fireground as soon as practical. Otherwise, serious injuries and fatalities will only continue to occur. Begin by taking full advantage of specific size-up factors that can indicate the presence of wind-driven fire conditions, as noted by NIST officials and experienced by many fire departments.

Size-up factors for high-rise, wind-driven fires

In 2008 NIST conducted testing at the scene of a seven-story structure on Governors Island, New York. The structure was vacant and once served as an apartment building. NIST research has identified key size-up factors that should be used by firefighters when responding to high-rise structures to assess possible wind-driven fire conditions.

In the seven-story structure experiments, it was determined that under conditions with a wind speed of 25 mph, the creation of a vent point could result in pressurizing fire venting from a side of a structure that could cause rapid fire spread with floor-to-ceiling and wall-to-wall involvement along an uninterrupted interior pathway. Exposure to these extreme fire conditions would not be survivable even for fully protected firefighters. It was specifically noted that one sign of possible wind-driven fire conditions could be as simple as observing smoke and fire attempting to push out into a corridor around a door's edges under force or through a burned-out door peephole as a way to determine if a window had been broken. This would indicate the need to use alternative tactics, because opening the door would result in inescapable and instantaneous wind-driven fire conditions that would be fatal.

Wind conditions at high-rise fires can be deceiving, as noted by Kerber: "There may be no wind on the first floor yet a significant wind condition on the 30th floor of the structure." Another sign of a wind-driven fire condition may involve a pulsing effect of fire venting out of a window against the wind, which is caused by an overpressurization of the room on fire (fig. 8–14) (Kerber and Madrzykowski 2009b).

Preventing Firefighter Disorientation

Fig. 8–14. A distinctive sign of a wind-driven fire condition is a fire pulse venting from a window, as seen here. (Courtesy of NIST.)

Note that the leaves on the tree in figure 8–17 have been forced over by the pressure of the wind, which serves as an indication of the need to implement tactics dedicated to wind-driven high-rise fires. Beforehand, however, an understanding of the importance of door control and thermal barriers to the spread of fire is necessary. Consult the NIST study as it pertains to the critical subject of door control during wind-driven fire experimentation.

Door control

The subject of door control was scientifically examined by Daniel Madrzykowski and Steve Kerber. The following is from Madrzykowski and Kerber's (2009) lab experiments:

> In Experiment 7 [Door Control], the fire was started with the door from the living room to the corridor in the closed position. The window failed at approximately 300 s. The door was opened at 377 s after ignition, this point is designated as time "zero" in Figure 6.2.3–1. Figure 6.2.3–1 clearly shows how the door was used as a WCD and a thermal barrier to protect the corridor from extreme thermal conditions. Temperatures along the flow

> path (corridor north position) exceeded 600 °C (1,112 °F) within 20 s of the door being opened. The temperatures in the south portions of the corridor, which were not in the flow path, increased at a much slower rate.

This valuable research enables firefighters to consider, before the fact, the consequences of initiating an interior attack during a wind-driven fire and to safely manage the possible flow path of fire that may dangerously vent out of points of entry. In general, rapid interior fire spread during wind-driven fires are associated with the following:

- A venting fire pressurized by a 10 mph or greater wind speed
- An unrestricted interior flow path (no interior door control)
- A downward or lower-pressure vent point

Recognizing these requirements and implementing tactics differently should help to prevent exposure to rapidly spreading fire. The following measures could additionally increase firefighter safety during wind-driven fires:

- Initiate interior door control by preplanning, that is, educating occupants before an incident occurs about the benefits of closing the door to the fire-involved unit and any other hallway doors whenever possible. The closed unit door in experiment 7 controlled the wind and protected the corridor by serving as a thermal barrier, a pressure barrier, and a barrier to oxygen flow into the fire area.

- Close interior doors when not attacking a fire. As shown by the 1999 Houston wind-driven high-rise incident, an attack on the unit on fire may have been safely initiated initially, but because of the need to exit, the apartment door was not closed. Also, because, as is customary during opened structure fires, the attack line remained in the involved room, inadvertently allowing an inward rush of wind from an open window to drive fire into a hallway due to an untimely change in wind speed and direction.

- Prevent the creation of an uninterrupted flow path and vent point by keeping stairwell and bulkhead doors closed. Typically, the point of firefighter entry to the fire floor in a high-rise incident is the stairwell. Coordinating the opening and closing of the bulkhead door and the stairwell door would prevent creating the dangerous intermediate flow path needed for fire to blow out of the fire floor, into the stairwell, and out any other opened doors above or below the level of the fire floor.

Doors, walls, and ceilings serving as wind control devices and thermal barriers

SOGs for wind-driven fires should include an understanding of the door control issue and logical assumptions regarding the presence or absence of doors and therefore their effectiveness during wind-driven fires. When the subject of door control is objectively examined with firefighter safety in mind, it is both the pressure and thermal barrier aspects of door control that are of utmost importance. In effect, keeping a door closed allows the door itself to function as a barrier to the spread of wind-driven fire along possible flow paths should a vent point exist or be created. In any structure, firefighters will encounter barriers against the rapid spread of fire caused by the wind in the form of solid or hollow interior and exterior doors.

If the interior doors are open during a wind-driven fire, there is no barrier and a flow path will exist. If the doors are closed, there is a barrier and a flow path will not exist as long as the doors remain intact. Once a door fails, an uninterrupted interior flow path will be created, and fire will engulf the space and rapidly spread in a floor-to-ceiling, wall-to-wall involvement through the flow path(s).

Walls that separate a room or area on fire also serve as barriers against the spread of fire as along as the walls remain intact. Ceilings that separate uninvolved rooms from involved attic spaces or ceiling spaces also function as barriers as long as the ceilings stay intact.

In the context of size-up and decision making during the course of a wind-driven structure fire, it is impossible for the first-arriving officer with only a view from the exterior to be able to predict if all possible barriers are open, closed, breached, or intact. Therefore, it is safest for the officer during wind-driven fires involving enclosed fire-resistive high-rise structures to make the assumption that interior barriers represented by walls or ceilings are intact, but that doors in fire-resistive construction may be opened or breached. This logic should be incorporated into SOG development for wind-driven fires in order to safely manage these extremely dangerous incidents.

High-rise wind-driven fire tactics

In summary, the high-rise wind-driven fire tactics described in this chapter, such as those evaluated by NIST, the Toledo Fire Department, the Chicago Fire Department, and the New York City Fire Department, should be utilized and incorporated into existing high-rise wind-driven fire SOGs. These tactics include the following:

As a first option:

1. Using appropriately powered and positioned positive pressure ventilators (PPVs) in the stairwell to keep the stairwell clear of smoke and heat.
2. Keeping stairwell and bulkhead or roof-access doors closed to prevent creating an uninterrupted flow path and vent point for smoke and fire.
3. Using WCDs to block the pressurizing wind and limit the oxygen reaching the fire, thus allowing crews standing by in the stairwell to initiate an interior attack on the room on fire. WCDs are made from flexible, fire-resistive materials.

As a second option:

1. In addition to numbers 1 and 2 above, if after WCD application, conditions on the fire floor remain untenable, initiate an attack on the seat of the fire from the exterior. When the fire floor is within reach of elevated master streams, these could be used to attack the fire as interior crews stand by in the stairwell. On higher fire floors, attacking the venting fire could be accomplished using specially designed high-rise nozzles from the window located directly below the window venting fire. For residential wind-driven fires, water flows of 100 to 200 gpm are sufficient to reduce the fire gas temperatures in the fire room and along the interior flow path.

2. For these commercial scenarios, high-rise master stream systems may be required. This nozzle system has been designed to flow at 625 gpm and 80 psi from the floor below. After knockdown has been accomplished from below, interior crews can safely advance down the hallway on the fire floor and enter the room of origin to complete final extinguishment.

Part of a 2008 study on wind-driven fires in high-rises conducted by the New York City Fire Department (FDNY) and NIST included the use of a high-rise nozzle to attack the fire from the floor below, as a means of keeping firefighters out of the potentially lethal flow path on the fire floor (figs. 8–15, 8–16, and 8–17).

Fig. 8–15. An FDNY firefighter positions the high-rise nozzle below the window of the involved apartment. (Courtesy of NIST.)

Fig. 8–16. As the line is charged, the nozzle directs the straight stream into the involved unit. (Courtesy of NIST.)

Fig. 8–17. Heavy conversion steam is produced as the stream strikes the ceiling of the involved room. (Courtesy of NIST.)

In the experiments, the use of the high-rise nozzle darkened down the fires and cooled down the corridor outside the apartment, which would then allow the attack, search, and ventilation operations to proceed effectively and safely on the fire floor. This is one of the attack options that has been added to the FDNY protocol for fire proof multiple dwelling unit fires (Kerber and Madrzykowski 2009a).

Figures 8–18, 8–19, and 8–20 highlight the dramatic effects of a 25 mph wind on a venting fire both without and with the assistance and benefit of a WCD. Testing was conducted by NIST in a vacant seven-story apartment building in 2008 on Governors Island, New York.

Fig. 8–18. As fire from a bedroom window attempts to vent against the air pressure created by the operating fan on the A side, a dangerous interior flow path of fire traveling down a hallway and venting out a living room window on the D side is created. (Courtesy of NIST.)

Fig. 8–19. After a WCD is placed completely over the pressurized window (the inlet), the wind needed for a flow of flames and heated fire gases to be pushed out the vent point is eliminated. However, if the fire has access to oxygen from the firefighters' entry point door, door control at that point will be needed, which would simultaneously allow implementation of a direct attack down the previously untenable corridor and living room. (Courtesy of NIST.)

Fig. 8–20. Placement of WCDs during the course of all wind-driven fires, whether involving grade or above-grade fires, provides a critical option in ensuring safer operations including prevention of engulfing fire exposure and sudden firefighter disorientation. In this photo, note the development of conversion steam indicating an engine crew has safely initiated a direct attack on the fire. (Courtesy of NIST.)

Preventing Firefighter Disorientation

Additionally, wind-driven fire conditions can materialize and endanger firefighters in all structures regardless of the height of a structure. For example, a wind-driven fire occurred and exposed firefighters on the 20th floor of an apartment high-rise in Chicago, which resulted in thermal burns to eight firefighters who were attempting a rescue from the room of origin; a 10th-floor fire of an apartment high-rise in New York City took the lives of three FDNY firefighters; and a 5th-floor fire of a condominium high-rise building in Houston, Texas, took the lives of one officer and one resident.

Fatal wind-driven fires have also taken place on grade level in residential structures in Prince William County, Virginia; Baytown, Texas; Houston, Texas; and in large, commercial, enclosed structure fires such as occurred in Wharton, Texas. These structure fires collectively took the lives of five firefighters attempting to make an evacuation of the structures. In all cases, an aggressive interior attack to locate and attack the seat of the fire or to conduct a primary search was initiated by the responding firefighters.

The need to effectively address several previously unrecognized fireground factors is required to prevent disorientation fatalities due to residential wind-driven structure fires. To specifically prevent exposure to fire that may lead to disorientation and loss of firefighter lives in residential structures on grade level requires training, awareness, and SOGs.

For example, as part of the standard size-up procedure, firefighters must remember that when certain conditions are encountered, relatively light winds may in fact rapidly move the fire toward advancing firefighters, which could lead to serious injuries or fatalities (fig. 8–21). However, through key actions taken by trained dispatchers and firefighters on the scene, the wind-driven fire hazard may be safely and effectively managed at the outset of every structure fire by members of every fire department in the country.

8   Avoiding Life-Threatening Hazards

Fig. 8–21. As 13 mph winds that gusted to 24 mph pressurized the B side of this mobile home, a West Virginia officer and firefighter died suddenly after advancing a 1½-in. handline 15 ft through the front doorway at the top of the steps. On arrival, conditions were that occupants had exited the structure and a nearby camper was involved in fire that had burned through a window on the B side, igniting the contents. Since hoods and facemasks were not worn prior to entering the home, the fallen firefighters may not have been fully aware of the extent of danger present during the incident. (Courtesy of West Virginia Fire Marshal, NIOSH F2009-07 [see NIOSH 2009d]).

Obscured vision and signs of breathing difficulty

There are conditions encountered at the scene of working structure fires where, due to high humidity and an absence of wind, smoke from a structure fire will blanket major portions of the fireground, obscuring the view of the structure and making breathing on the exterior of the structure difficult. These conditions create challenges for first-arriving firefighters because obviously size-up and command requires the ability to see the structure. In spite of these dangerous conditions, strategies and tactics must nonetheless be safely and effectively implemented.

Exterior signs observed specifically during windy conditions may consist of a mass of heavy, spreading, ground-hugging smoke, commonly from the A side, causing an inability to see the structure and a general inability to breathe when positioned downwind from the structure. These conditions may signal that firefighters have arrived on the downwind side of a wind-driven fire and should be interpreted as a warning that firefighters will be dealing with an extremely dangerous structure fire that calls for use of nontraditional strategies and tactics that incorporate the findings about wind-driven fire behavior.

The Oak Park, Michigan, incident

Off-duty firefighter and photographer Dennis Walus provided the following fire scene information:

> On Wednesday June 2, 2010, at approximately 5:00 p.m., The Oak Park Department of Public Safety responded to a report of smoke in the building at 23111 Coolidge and 9 Mile Road [fig. 8–22]. Upon arrival, the first officer on the scene reported heavy smoke showing from a 250 ft. × 150 ft. occupied commercial building. Oak Park Public Safety prepared for an interior fire attack. While attempting an attack on the fire, the interior ceilings started to collapse. Incident command then ordered all crews to stay out of the building and started a defensive attack on this fire. Oak Park DPS requested off-duty callbacks and requested mutual aid from Beverly Hills Public Safety, the Ferndale Fire Department, and the Troy Fire Department. A ladder pipe and several handlines and ground monitors were put into operation. The occupied building had roof collapse and was a total loss. Winds were from the west at 17 mph gusting to 32 mph during this large enclosed commercial structure fire. (Email correspondence June 3, 2010)

Fig. 8–22. Oak Park, Michigan, firefighters encountered obscured vision on the exterior of a large, occupied, enclosed commercial structure. (Courtesy of Dennis Walus.)

The obscured vision sign or size-up factor was present during several enclosed structure fires in the past. Unfortunately, many resulted in fatalities when a traditional attack from the unburned side and through the structure's front doorway (the dangerous vent point side) was quickly initiated on arrival.

The angle of the smoke plume

To determine if wind-driven fire conditions actually exist at an early stage of a structure fire, it is very important for firefighters to know the wind speed. Although weather stations may provide an official wind speed reading, the reading may not be the same at the scene of a structure fire, which may be located several miles away from the weather station. Therefore a backup means of making a calculation is needed.

Approximate wind speed can be logically estimated at the start of structure fires. This can be done on the basis of reported wind speeds noted in conjunction with the corresponding angle of smoke plumes photographed and observed during past structure fires.

Whenever a smoke plume venting from a structure fire rises straight above the structure in a column, the smoke can be described as calm and not associated with an appreciable wind speed. However, and important to the safety of firefighters, whenever wind causes a smoke plume to form at a 45° angle to the structure, the corresponding wind speed is approximately 12–15 mph. In addition, whenever the wind causes an almost horizontal smoke plume (90° angle), the corresponding wind speed is approximately 30 mph or greater.

When considered in conjunction with wind speed data from NIST, which noted a wind speed as low as 10 mph could cause a wind-driven fire, a smoke plume angle that is tilted to a slightly less than 45° angle (45° corresponds to 12–15 mph wind speed) would therefore be considered adequate to produce a dangerous wind-driven fire.

This important wind speed sign, represented by the angle of a smoke plume seen from any structure fire, may help officers and firefighters to quickly estimate the wind speed at the scene and possibly avoid exposure to disorienting fire caused by a wind-driven fire condition. In summary, a smoke plume angle of slightly less than 45° is approximately equal to a 10 mph wind speed; a smoke plume angle of 45° is approximately equal to a 12–15 mph wind speed; and a smoke plume angle of 90° is approximate to a 30 mph or greater wind speed. Firefighters can be easily deceived by how low a breeze can actually produce a hazardous wind-driven fire, and they often feel that a much higher speed of perhaps 20 mph is needed. However, according to NIST, wind speeds as low as 10 mph are enough to cause extremely dangerous wind-driven fires. Objectively speaking, that is only a slight breeze.

The following information and photographs (figs. 8–23, 8–24, and 8–25) illustrate and summarize how smoke plume angles and corresponding wind speeds can be used to quickly determine if wind-driven fire conditions actually exist at the scene of a structure fire.

Fig. 8–23. In a 2001 incident involving a Phoenix supermarket fire, winds of 15 mph pressurized the B side of the building. During this incident, heavy smoke also obscured the visibility at the command post on the A side. As an informal size-up rule, a 45° angle seen along the upper surface of a smoke plume indicates a 12–15 mph wind speed, which is above the 10 mph speed required for a structure fire, under certain conditions, to be considered wind-driven, thus requiring strategies and tactics better suited for managing the wind factor. (Courtesy of Paul Ramirez.)

Preventing Firefighter Disorientation

Fig. 8–24. Winds of 17 mph gusting to 32 mph pressurized the C side of this commercial structure in Oak Park, Michigan, hampering firefighters as the winds accelerated fire conditions, rapidly destroying the structure. Corresponding to a 17 mph wind speed, the upper surface of the smoke plume in the photo indicates an angle slightly greater than 45°. This heavy, ground-hugging smoke will also obscure visibility and create breathing difficulties for firefighters at similar types of structure fires. (Courtesy of Dennis Walus.)

Fig. 8–25. The 90° angle along the upper surface of the smoke plume is indicative of the force of a 32 mph wind gust reported and experienced by firefighters at the scene. (Courtesy of Dennis Walus.)

8 — Avoiding Life-Threatening Hazards

The wind-driven fire action plan

To avoid firefighter disorientation and fatalities caused by wind-driven fires, firefighters should consider the following wind-driven fire action plan. Execution of this plan is demonstrated at the scene of a vacant residence with an opened structure and a basement by members of the Pratt and Wichita Fire Departments (both in Kansas). The following series of photographs (figs. 8–26, 8–27, 8–28, 8–29, 8–30, 8–31, and 8–32) provide a view of the action.

Fig. 8–26. This is a view of the A-D corner of a vacant, involved opened structure with a basement. (Courtesy of Andy Thomas.)

1. Firefighters must be trained to understand that a wind speed as low as 10 mph or greater pressurizing venting fire on a side of a structure can cause sudden life-threatening fire conditions on the interior if a vent point and flow path are created.

2. As a warning and reminder to consider a wind-driven fire condition with a 10 mph or greater wind speed, dispatchers must transmit the wind speed and direction to responding companies at the time of dispatch. Command must also be notified of any forecasted change in the wind speed and direction during the incident and make tactical changes accordingly.

Preventing Firefighter Disorientation

Fig. 8–27. Venting fire located at the B-C corner. (Courtesy of Andy Thomas.)

3. A complete 360° walk-around should be conducted to determine if vented fire is being pressurized by the wind and, if so, on which side of the structure, including the roof and basement.

4. When a wind-driven fire condition is encountered, all responding companies must be notified either by radio transmission or in a face-to-face exchange.

Fig. 8–28. Firefighters safely attack the fire on the C side with the wind at their backs. (Courtesy of Andy Thomas.)

Fig. 8–29. Firefighters move forward while continuing to apply water. (Courtesy of Andy Thomas.)

5. From the exterior, engine companies should quickly attack the fire on the pressurized side of the structure to knock down the main body of fire. When this is accomplished and if the building is structurally sound, crews may enter through the extinguished side to conduct a primary search as other firefighters advance to check for fire extension. In the absence of an engine company, truck companies equipped with fire-resistant fire curtains, when first to arrive on the scene, may consider suspending fire curtains from ground ladders to seal the openings and then safely initiate a primary search.

Fig. 8–30. Firefighters inspect the basement level with the assistance of a thermal imager. (Courtesy of Andy Thomas.)

Fig. 8–31. Firefighters pull the ceiling immediately inside the window to check for fire extension. (Courtesy of Andy Thomas.)

Fig. 8–32. A positive pressure ventilator (PPV) is set up to assist in ventilation, while a chief officer determines fire cause and extent of property damage. (Courtesy of Andy Thomas.)

In the development of a wind-driven fire SOG, department leadership, including the chief, training, and safety officers, may wish to determine when the wind speed and direction should be transmitted by dispatchers during reports of working structures fires. For example, the wind-driven fire SOG may stipulate that whenever the prevailing wind speed or wind gusts reaches or exceeds 8 mph, for a margin of safety, dispatch will transmit the wind speed and direction. There may also be unusual circumstances during wind-driven fire situations that firefighters should be aware of and factored into their size-up. For example, due to variations in the elevation, location, and height of adjacent structures that may shield the structure fire from the wind, fire venting from a structure may not be affected by the pressurizing effects of the wind and therefore use of the NIST-UL method of initial attack from the exterior prior to initiating an interior attack may be considered. The 360° size-up conducted by a trained officer or firefighter will confirm this condition, and the first-arriving officer in command should therefore clearly note in the initial fire report that the wind is or is not a factor. There may also be situations where the monitoring weather station from which dispatchers obtain wind speed and directional information may be located several miles away from the location of the structure fire, and the wind speed at the scene in fact may be higher or lower or from a different direction than initially reported. Sound officer and command judgment therefore should always be exercised.

Preventing Firefighter Disorientation

Rapid fire spread from above

Firefighters should be aware that if a vent point and unimpeded flow paths are established, wind-driven fires may engulf firefighters operating in either high-rise buildings or structures on grade level. Additionally, for safety and survival, firefighters must know that wind can also force fire into overhead attic spaces, ultimately threatening interior firefighters working below.

In these cases, the fire may enter the attic space as wind pressurizes the fire originating from the exterior walls or as a result of vertical ventilation. If, during a wind-driven attic fire, the ceiling is pulled by interior truck crews or fire burns through the ceiling, which serves as an overhead thermal barrier, a downward flow of wind-driven fire from the ceiling space above will suddenly engulf interior firefighters. The following photographs (figs. 8–33, 8–34, 8–35, 8–36, and 8–37) involving an automatic aid response of the Arkansas City and Winfield, Kansas, Fire Departments are of a well-managed, hazardous wind-driven attic fire.

Fig. 8–33. The wind at this 2009 Arkansas City, Kansas, incident pressurized the roof from the A side and engulfed the attic space at this unoccupied two-story residence. (Courtesy of Jason Denny.)

8 — Avoiding Life-Threatening Hazards

Fig. 8–34. Within minutes, the fire rapidly spread downward past unobstructed flow paths and vented out of the first floor on the C side. (Courtesy of Jason Denny.)

Fig. 8–35. Note the angle of the smoke plume and the heavy ground-hugging smoke, indicative of a wind-driven fire. (Courtesy of Jason Denny.)

Fig. 8–36. Fire was venting out of the entire C side of the structure in this photo. This is also a sign of an extremely dangerous wind-driven fire calling for suitable dedicated tactics. (Courtesy of Jason Denny.)

Fig. 8–37. While keeping the wind at their backs, Winfield, Kansas, firefighters who assisted at the incident safely and effectively used elevated master streams to attack the fire from the A side of the structure. (Courtesy of Jason Denny.)

Rapid downward fire spread that resulted in a line-of-duty death occurred during a wind-driven residential fire that took place in Prince William County, Virginia, in 2007. During the incident Prince William County firefighters found fire and smoke showing on the B-C corner of a residence. As a northwest wind of 25 mph gusting to 48 mph pressurized the C side, a primary search was initiated through the front door on the A side. In the second-floor master bedroom, light smoke at the ceiling suddenly changed to heavy black smoke and orange flames that disoriented and trapped one firefighter during this wind-driven residential fire. Unknown to the responders, the occupants had exited the structure prior to firefighter arrival (NIOSH 2008).

Rapid fire spread from below

Firefighters must be aware that under the right circumstances, wind speeds as low as 10 mph pressurizing fire venting from a structure can rapidly spread through the structure on or above grade levels. However, a small house fire involving a basement on February 24, 2012, in Prince George's County, Maryland, also showed that wind-driven fires can rapidly spread from below-grade levels and travel up and out on grade levels if an unrestricted interior flow path and vent point exist.

Seven Prince George's County firefighters were engulfed and injured by this wind-driven fire. According to the fire chief, winds were gusting out of the west up to 40–45 mph as they attempted to initiate an interior attack on a basement fire from the front door. The winds, however, were blowing directly at—and into—the burning basement, which had a west-facing door. According to the chief, "As soon as the guys opened the front door and advanced, it blew from the basement, up the steps, and right out the front door." The chief continued, "It was like a blowtorch coming up the steps and right out the door." The entire incident, "from the time they were in the door until they were burned—took eight seconds" (du Lac 2012).

Disorienting hazards associated with high-rise fires

Of the 17 cases examined during the disorientation study (Mora 2003), three involved firefighter disorientation within high-rise hallways. In New York City, three firefighters became disoriented after sudden exposure to a wind-driven fire that caused their fatalities. In another, similar high-rise incident in Houston, Texas, a wind-driven fire contributed to the disorientation and fatality of a veteran fire officer as he unsuccessfully attempted to evacuate the fire floor with another officer. In an incident in St. Louis, Missouri, two firefighters became disoriented in a hallway charged with thick blinding smoke as they conducted a primary search, resulting in career-ending injuries.

Analysis of disorientation cases that occurred in high-rise hallways found that in two incidents, a handline, meant for the protection of firefighters and to serve as a lifeline to exit the floor, was not used. In one case, a handline was used to attack the fire, but because it remained in an apartment unit as firefighters evacuated, it allowed the door to remain ajar and wind-driven fire to subsequently enter the smoke-filled hallway. In all cases, handline separation or entangled handlines were encountered, leading to a loss of company integrity, which serves as a fire company's all-important safety net.

Four specific hazards are known to contribute to the disorientation of firefighters during high-rise incidents:

- Prolonged zero visibility conditions (PZVCs)
- Handline separation during PZVCs
- Entangled handlines encountered during PZVCs
- Exposure to wind-driven fire

Safety measures can be taken before and during high-rise fires to prevent firefighter disorientation. The measures that may be incorporated into departmental high-rise SOGs include the following:

- Routine use of fire-resistive WCDs to seal the wind trying to enter a structure's window
- Maintaining company integrity

- Maintaining contact with a handline
- Routine use of advanced photoluminescence technology to increase firefighter and equipment visibility in darkened environments
- Routine use of safety directional arrows on attack lines
- Routine use of thermal imaging cameras
- Pressurizing a stairwell and fire floor with strategically placed and adequately powered PPVs to control smoke in the stairwell and significantly reduce interior temperatures

During wind-driven fire conditions or when heavy fire is showing from multiple windows, it may be too dangerous to send firefighters to the unit or units on fire. As an alternative, when the fire is accessible from the exterior by vehicle-mounted or elevated master streams, and life safety is not an issue, coordinated use of these master streams should be initiated to quickly knock down the bulk of the fire. Remaining fire extension or hot spots can then be extinguished from the interior.

Connecting attack lines to the standpipe on the floor below the fire instead of on the fire floor itself can also help to avoid exposure to life-threatening hazards that contribute to disorientation and serious firefighter injuries or fatalities.

During wind-driven fire conditions, a significant barrier to rapid fire spread is lost if a door to an involved high-rise room is open or has burned through. During these scenarios, fire venting from windows of an involved high-rise, as wind pressurizes the venting fire, represents an extremely dangerous condition for interior firefighters. If integrity of the door is lost, the wind will push or drive fire to the point of entry, typically the stairwell doorway that advancing firefighters would use to reach the floor and involved room.

As previously noted, New York City firefighters have designed nozzles used specifically to fight high-rise fires from the safety of the floor below the fire during wind-driven fire conditions.

Another option available to fight fires during wind-driven apartment high-rise fires involves the use of penetrating nozzles from safe positions on the floor directly above the unit on fire. It is conceivable that multiple penetrating nozzles, each rated at 175 gpm, could be used simultaneously to attack the fire. However, command must initially consider the safety of working above the fire before using this option.

Rollover

The Atlanta rollover incident

In 2006, Atlanta, Georgia, firefighters found heavy smoke showing from a small enclosed structure involving an abandoned duplex residence (fig. 8–38).

Fig. 8–38. This is the A side of the structure. The entire duplex, which measured 36 ft. wide and 34 ft. deep, had windows and doors boarded with plywood. Each duplex had 500 square feet of living area, and the structure was divided along the center. (Courtesy of NIOSH F2007-02 [see NIOSH 2009a].)

8   Avoiding Life-Threatening Hazards

After advancing through the front door, a flameover (rollover) occurred with a subsequent flashover fully engulfing the three-member attack crew and other firefighters. Two firefighters on the attack crew as well as other disoriented firefighters were able to evacuate through the front door. However, the third firefighter, also disoriented by the blinding flashover, was seen running through the flames by the front entrance and did not exit. He was later located along the interior D wall of the structure and quickly removed from the structure, but did not survive his injuries (NIOSH 2009a).

To summarize, clearly there are certain situations on the fireground that are life-threatening and may benefit from survival contingencies that do not currently exist at the company level. Sudden exposure to rollover, flashover, and ventilation-induced flashover are three of the most dangerous. However, lessons have been learned. During these hazardous situations, attack rules must be established and understood by members of individual fire companies before the fire and implemented on the fireground when needed and without hesitation. These survival rules are simple and can be successfully integrated and used to ultimately help prevent the disorientation and injury of firefighters:

- Do not enter rooms about to flashover.
- Exit rooms about to flashover.
- Attack the seen or unseen fire during rollover, flashover, or vented room flashover scenarios.

Advancing firefighters must therefore not only look for signs of imminent flashover that may mushroom unseen overhead, but they must also include contingencies that call for immediately and fully opening nozzles set to semifog or full fog patterns while aiming the stream directly ahead to provide protection against an unexpected fast approaching wall of fire.

The value of routinely having quickly deployed backup streams for unanticipated fire spread that may trap the first-advancing company and having multiple and maneuverable backup lines discharging flows that approach or equal 200 gpm cannot be overemphasized, especially during the types of life-threatening situations discussed here.

Floor Collapse

Life-threatening events, such as the collapse of a fire-weakened floor, produce firefighter disorientation that can result in critical injuries and deaths. Collapses of fire-weakened floors have repeatedly caused the fatalities of action-oriented firefighters who may not have associated basement fires with an extreme degree of danger. Even though there have been instances of firefighter survival following first-floor collapses, these are generally extremely dangerous situations.

First-floor collapses caused by fire exposure, as in the case of roof collapses, are typically associated with the immediate production of heavy smoke and fire, which often precludes any rescue attempt until after the bulk of the fire is knocked down. This effort is often much too late to achieve firefighter or occupant survival. In addition, these tragic events often occur because the extreme hazard of a potential floor collapse may not have been apparent or even considered by responders. Possibly due to a misinterpretation of only light to moderate smoke conditions found on arrival, which may have given the appearance of a "food on the stove" call, a casual walk into the structure to investigate may be made, with disastrous consequences.

Another reason the collapse potential may be overlooked deals with the tunnel vision that is caused by thinking that a primary search is genuinely needed following a quick response to the scene. In these scenarios, firefighters may incorrectly assume that the structural integrity of first floors in opened structures with basements have not been compromised enough to prohibit walking on top of the fire-weakened first floors (fig. 8–39). Every precautionary measure must be taken by command and responders to avoid falling into involved basement fires or being exposed to the multiple life-threatening hazards associated with making a fast and aggressive interior attack into a basement fire.

Fig. 8–39. This is an unoccupied, residential opened structure with a basement, which took the life of a volunteer mutual aid firefighter. (Courtesy of NIOSH F2008-26 [see NIOSH 2009e].)

Prolonged Zero-Visibility Conditions

Prolonged zero visibility conditions (PZVCs) are extremely dangerous because they persist longer than the effective 15-minute breathing time offered by the commonly used 30-minute rated SCBA, (fig. 8–40).

Fig. 8–40. The lack of adequate means of egress for escape and ventilation at enclosed structure fires, such as the Charleston incident, contribute to life-threatening hazards. The most common hazard is disorientation secondary to PZVCs. (Courtesy of Stewart English.)

For safety and to take into account emerging SCBA technology, an expanded definition for PZVCs should be any zero visibility condition that will meet or exceed any effective breathing time *regardless of the rated duration of the breathing apparatus.* Therefore, PZVCs also pertain to those conditions that exceed SCBA's 20-minute, 1-hour, 2-hour, or longer effective breathing times. Additionally, regardless of the reason, should handline separation or entangled handlines be encountered during PZVCs, disorientation, injuries, and fatalities due to inhalation of smoke or superheated gases may result.

Aside from using thermal imaging cameras (within limitations), there are two additional ways firefighters can safely manage PZVCs:

- Development and use of technology that offers firefighters SCBA of equal or lighter weight than the SCBA tank currently used while providing significantly longer breathing times.

6. Use of *short interior attacks* that reduce or minimize exposure time by simply reducing or minimizing the travel distance between the exterior and the seat of the fire. This key enclosed structure tactic utilizes forcible entry or the breaching of exterior or interior adjoining walls located close to the seat of the fire to provide firefighters with reduced entry and exit travel times (Mora 2003).

The ideal planned response to enclosed structure fires should ultimately allow firefighters to respond in adequate numbers, with the best of all available technology including tactical and safety contingencies that avoid disorienting, life-threatening hazards during any enclosed structure fire.

As mentioned, these may come in the form of full personal protective equipment, thermal imaging cameras, lighter and longer-lasting SCBAs, simple wire cutters, safety directional arrows for handlines, safety directional compasses, reliable portable radios for all firefighters, fire curtains, and advanced photoluminescence technology (which illuminates firefighters and equipment in dark environments), and presence of certified safety officers and rapid intervention teams. The use of fast and aggressive interior attacks should be avoided without first carefully considering overall safety and effectiveness following a cautious interior assessment of conditions. And finally use of a *short interior attack* to avoid multiple life-threatening hazards, including PZVCs, common to enclosed structure fires should be an integral part of the approach to safely manage enclosed structure fires.

Preplanning Unprotected Enclosed Structures

Although the potential of enclosed structure incidents to occur in any community is tremendous, they may be safely managed at the local level with appropriate staffing, training, technology, and preplanning procedures. Furthermore, this hazard can be effectively and

Preventing Firefighter Disorientation

immediately managed by knowing where in the response area the unprotected enclosed structures are located.

In this effort, fire companies should tour their districts to locate enclosed structures that are not protected by automatic sprinkler systems (fig. 8–41). The completed address list should then be forwarded to the dispatch office so that firefighters can be forewarned about the hazardous nature of the structure prior to responding, should a fire break out at any of those locations.

Fig. 8–41. To help ensure the safety of all firefighters who may respond to a structure fire, officers can develop, during the course of preplan activities, an address list of unprotected enclosed structures located in their first-due area. The completed list would then be forwarded to the dispatch office for future reference. During a report of a structure fire whose address appears on the extremely dangerous structure list, dispatchers can then alert responding companies verbally or by means of mobile data terminals that the structure fire they are responding to is in fact an extremely dangerous unprotected enclosed structure. With this advance warning, arriving firefighters will know that enclosed structure tactics will need to be used to help ensure a safer and more effective operation.

By preplanning and through understanding of the hazards during multicompany settings, firefighters can prepare for an incident involving these structures by using enclosed structure tactics or an enclosed structure SOG programmed to avoid the risks. In addition, a more long-term solution may involve the installation of automatic sprinkler systems in these unprotected enclosed structures and ensuring they are properly maintained.

Additionally, the value of safe, flexible, and effective hose evolutions, which represent a major component in the prevention of firefighter disorientation and coordinated ventilation during the course of an opened or enclosed structure fire, cannot be overemphasized. Therefore, prior to discussing the aspects of enclosed structure tactics that would be executed sequentially based on company or resource arrival, the next two chapters cover the specific hose evolutions in the context of the new enclosed structure approach. A review of the importance of providing coordinated ventilation is also included.

References

Barowy, A., & Madrzykowski, D. 2012a. "Analysis of a Fatal Wind-Driven Fire in a Single-Story House." *Fire Engineering*, vol. 165, no. 6, 63–74

———. 2012b. *Simulation of the Dynamics of a Wind-Driven Fire in a Ranch-Style House—Texas (NIST TN 1729)*. Gaithersburg, MD: National Institute of Standards and Technology.

du Lac, J. F. 2012. "Blaze That Injured 7 Prince George's Firefighters Called 'Freak Occurrence'." *Washington Post*, February 25. https://www.washingtonpost.com/local/blaze-that-injured-7-prince-georges-firefighters-called-freak-occurrence/2012/02/25/gIQAdGJMaR_story.html.

Gorbett, G. E., and Hopkins, R. 2007. "The Current Knowledge & Training Regarding Backdraft, Flashover, and Other Rapid Fire Progression Phenomena." Presentation at the National Fire Protection Association World Safety Conference Boston, Massachusetts, June 4, 2007. http://Tracefireandsafety.com/FireInvestigationResearch/Flashover%20Backdraft%20paper-NFPA%20Boston.pdf.

International Association of Fire Chiefs (IAFC). 2012. "Rules of Engagement for Structural Firefighting: Increasing Firefighter Survival." Developed by the Safety, Health and Survival Section International Association of Fire

Chiefs. http://websites.firecompanies.com/iafcsafety/files/2013/10/Rules_of_Engagement_short_v10_2.12.pdf.

International Fire Service Training Association (IFSTA). 2008. "Fire Behavior." In *Essentials of Fire Fighting,* 5th ed. Stillwater, OK: Fire Protection Publications, Oklahoma State University, 122.

Kerber, S., and Madrzykowski, D. 2009a. *Fire Fighting Tactics Under Wind Driven Conditions: 7-Story Building Experiments* (Technical Note No. 1629). Gaithersburg, MD: National Institute of Standards and Technology.

———. 2009b. *Evaluating Fire Fighting Tactics Under Wind Driven Conditions.* Training DVD, disc 1. Gaithersburg, MD: National Institute of Standards and Technology.

Kerber, S. 2013. "What Research Tells Us About the Modern Fireground." *FireRescue.* http://www.firefighternation.com/article/strategy-and-tactics/what-research-tells-us-about-modern-fireground.

Madrzykowski, D. 2013. "Fire Dynamics: The Science of Fire Fighting." *International Fire Service Journal of Leadership and Management*, vol. 7. Stillwater, OK: Fire Protection Publications/IFSTA, 71–72.

Madrzykowski. D., and Kent, J. L. 2011. *Examination of the Thermal Conditions of a Wood Floor Assembly Above a Compartment Fire* (Technical Report NIST TN-1709). Gaithersburg, MD: National Institute of Standards and Technology.

Madrzykowski, D., and Kerber, S. 2009. *Fire Fighting Under Wind Driven Conditions: Laboratory Experiments* (Technical Note 1618). Gaithersburg, MD: National Institute of Standards and Technology.

———. 2010. "Wind Driven Fire Research: Hazards and Tactics." *Fire Engineering*, vol. 163, no 3, 79–94.

Mora, W. R. 2003. "U.S. Firefighter Disorientation Study 1979–2001." http://www.sustainable-design.ie/fire/USA-San-Antonio_Firefighter-Disorientation-Study_July-2003.pdf.

National Fallen Firefighters Foundation (NFFF). 2011. "Firefighter Life Safety Initiatives." http://www.lifesafetyinitiatives.com.

National Fire Protection Association (NFPA). 2013a. "NFPA 555: Guide on Methods for Evaluating Potential for Room Flashover." http://www.nfpa.org/codes-and-standards/document-information-pages?mode=code&code=555.

———. 2013b. "NFPA 1971: Standard on Protective Ensembles for Structural Fire Fighting and Proximity Fire Fighting." http://www.nfpa.org/codes-and-standards/document-information-pages?mode=code&code=1971.

National Institute for Occupational Safety and Health (NIOSH). 1998a. "Backdraft in Commercial Building Claims the Lives of Two Fire Fighters, Injures Three, and Five Fire Fighters Barely Escape—Illinois." http://www.cdc.gov/niosh/fire/reports/face9805.html.

———. 1998b. "Single Family Dwelling Fire Claims the Lives of Two Volunteer Fire Fighters—Ohio." http://www.cdc.gov/niosh/fire/reports/face9806.html.

———. 2004. "Partial Roof Collapse in Commercial Structure Fire Claims the Lives of Two Career Fire Fighters—Tennessee." http://www.cdc.gov/niosh/fire/reports/face200318.html.

———. 2006. "Career Battalion Chief and Master Fire Fighter Die and Twenty-Nine Career Fire Fighters are Injured during a Five Alarm Church Fire—Pennsylvania." http://www.cdc.gov/niosh/fire/reports/face200417.html.

———. 2008. "Career Fire Fighter Dies in Wind Driven Residential Structure Fire—Virginia." http://www.cdc.gov/niosh/fire/reports/face200712.html.

———. 2009a. "Career Fire Fighter Injured during Rapid Fire Progression in an Abandoned Structure Dies Six Days Later—Georgia." http://www.cdc.gov/niosh/fire/reports/face200702.html.

———. 2009b. "Nine Fire Fighters from a Combination Department Injured in an Explosion at a Restaurant Fire—Colorado." http://www.cdc.gov/niosh/fire/reports/face200803.html.

———. 2009c. "Two Career Firefighters Die While Making Initial Attack on a Restaurant Fire—Massachusetts." http://www.cdc.gov/niosh/fire/reports/face200732.html.

———. 2009d. "Volunteer Lieutenant and a Fire Fighter Die While Combating a Mobile Home Fire—West Virginia." http://www.cdc.gov/niosh/fire/reports/face200907.html.

———. 2009e. "A Volunteer Mutual Aid Fire Fighter Dies in a Floor Collapse in a Residential Basement Fire—Illinois." http://www.cdc.gov/niosh/fire/reports/face200826.html.

9
HOSE EVOLUTIONS AND FIREGROUND REALITIES

The ability to prevent firefighter disorientation, accomplish successful rescues, and minimize property loss to the greatest degree possible is associated with adequate staffing, training, and the time required performing an effective attack during the initial stage of a structure fire. Achieving these objectives, however, requires the use of safe and effective hose evolutions that will allow firefighting tasks to be accomplished as quickly as possible to maximize safety on the fireground. Unfortunately, too many traumatic firefighter and civilian fatalities still take place at the scene of structure fires.

When these tragic events are closely examined, the circumstances surrounding many of the fatalities may at first appear to be unpreventable. In some cases, an engulfing flashover may have occurred while first-arriving firefighters were engaged in attacking or locating the seat of the fire. In other cases, firefighters and civilians may have perished as a result of deadly flashovers during primary search or rescues.

As a result of these unfortunate fires, recommendations to help prevent similar outcomes have been offered, yet flashover fatalities still take place. Furthermore, it is not unreasonable to anticipate tragic outcomes to continue if firefighters continue to attack structure fires the same way. If any reduction in firefighter and civilian fatalities in the United States is to be achieved, there must be improvement in the effectiveness of the initial and ongoing fire attack, based on the latest information available concerning the degrees of risk associated with opened and enclosed structures fires and on the correct interpretation of the initial size-up factors during structure fires.

Additionally, solutions offered here are based on the use of hose evolutions that will work best with the resources available in each department, the use of dual pumping at the scene, and provision of flexibility based on conditions encountered.

Furthermore, certain fireground problems or realities often appear during the course of a working fire. For example, it is true that many structure fires have escalated to a greater alarm because of an initial shortage of water at the scene. Other examples may have involved structure fires where operator or mechanical difficulties were encountered by one of the first-arriving engines. Due to possible unfavorable operational outcomes, these potential problems and others must be effectively addressed by incorporating suitable changes into departmental standard operating guidelines (SOGs) to help prevent future firefighter injuries and fatalities.

The primary objective of the first-arriving company during every structure fire is to ensure that everyone is out of the structure. If it is thought that occupants are in the structure, the first step is to quickly and safely initiate a primary search. When occupants are found, firefighters then focus on quickly removing them from the structure. Success in this endeavor is associated with the number of firefighters, the number of engines and trucks, and the amount of water available to simultaneously attack the fire while a search by other firefighters is conducted.

The fastest and most effective way to accomplish this goal is to lay and advance charged preconnected handlines from different engines as soon as possible, position them between the fire and the victims, and aggressively apply an adequate amount of water onto the fire from more than one position. This approach, which advocates early simultaneous action taken within understood limitations, accomplishes five objectives:

1. It allows the fire to be quickly knocked down to avoid the risk of collapse.
2. It allows the fire to be quickly knocked down to avoid the prolonged zero visibility conditions (PZVCs).

9

Hose Evolutions and Fireground Realities

3. It protects the means of egress.
4. It stops the damage to property in a more timely fashion.
5. It prevents flashover, thus allowing firefighters the time needed to safely and simultaneously initiate a primary search and to remove victims from the structure at the earliest stage of a fire.

By accomplishing the fifth objective, the victims and the firefighters are both protected from hazards such as fire engulfing the point of entry; burned handlines, which lose pressure and become unusable; and untenable heat. These benefits are realized when enough water is available to simultaneously cool hot interiors with handlines as they are advanced to hit the seat of the fire as well as other areas of fire involvement.

Effective and coordinated ventilation of smoke and heat from the structure is also needed in order to increase visibility and again reduce the chance of a fatal flashover. On numerous occasions, however, and when every effort humanly possible was attempted to help trapped occupants, firefighters have become disoriented following blinding flashovers that prevented their safe evacuation from the involved structure.

Tragic scenarios resulting in firefighter disorientation and following flashover, collapse, PZVCs, or conversion steam produced at any structure, opened or enclosed, however, may be prevented in the future by the use of faster developing hose evolutions that more effectively address these specific hazards earlier in the incident.

In the effort to achieve a quick, safe, and effective fire attack, there are several tactical approaches currently in use across the country. However, when the tactics are objectively examined, all are adequate, but some actually are more effective than others at minimizing property loss and maximizing the safety of civilians and firefighters.

With the ultimate goal of preventing the disorientation and fatality of firefighters, the following sections provide a review of traditional and contemporary approaches to fire flow development, including benefits and drawbacks for firefighter and departmental consideration.

Traditional Approaches to Fire Flow Development

The following information assumes that a department works within a community protected by a water supply system that follows the American Water Works Association (AWWA) guidelines and is approved to serve as a public water resource. These guidelines include adequate and reliable water flow for both private and fire flow requirements; elevated water tanks; pumps of sufficient number, volume, and capacity; and an adequate water distribution system consisting of mains of adequate diameter to serve the needs of the various areas of the community. The guidelines also require the distribution and spacing of hydrants to protect structures in residential and commercial districts (AWWA 1999).

With that in mind, the examples of 5-in. hose evolutions shown next provide responding companies with the choices and therefore the flexibility to achieve the best possible outcome on the fireground. Evolution implementation should follow nationally recognized standards and practices such as use of the Occupational Safety and Health Administration's two-in/two-out rule, rapid intervention teams, safety officers, and the Incident Command System, as noted in *NFPA 1500: Standard on Fire Department Occupational Safety and Health Program*, is assumed.

The attack-supply evolution (blitz attack)

One of most commonly used hose evolutions to help accomplish the tactical priorities of rescue, fire control, and conservation of personal property is the attack-supply evolution, also referred to as the blitz attack, (fig. 9–1).

As the phrase implies, in this evolution the first-arriving engine company immediately lays and charges a preconnected 1¾-in. handline and begins an interior attack, using the water from the booster tank. While the first company attacks, the second arriving engine lays a 5-in. supply line to establish a continuous supply of water for the engine operating on the scene.

9   Hose Evolutions and Fireground Realities

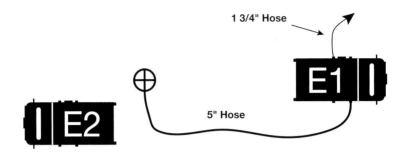

Fig. 9–1. This traditional evolution—the attack-supply evolution, also referred to as the blitz attack—involves an attack by the first-arriving engine company, initially using water from the booster tank, while the second engine supplies the first.

This evolution works relatively well, but on closer examination, it has some weaknesses. When using this evolution and during the early stage of a structure fire, firefighters are keenly aware of the fire flow limitation. During this time frame, firefighters clearly understand that the only water immediately available to fight the fire and to protect the advancing firefighters and victims in the structure is the volume of water carried in the booster tank of the first-arriving engine.

Booster tanks typically hold either 500 or 750 gallons of water, though there are certain engines equipped with 1,000- or 1,500-gallon booster tanks. This initial volume limitation is clearly understood by all firefighters, which is the reason why when first on the scene, firefighters use tank water reasonably and judiciously when making an interior attack, while always being prepared to exit the structure if necessary during this early and dangerous firefighting time frame.

It is only when the interior firefighters hear the approaching sirens, the radio transmission announcing that the second-due engine has arrived on the scene, and ultimately the sounds of a second working fire stream that they will exercise a greater amount of aggressiveness with their handline and remaining amount of booster water.

A second common drawback is that initially there is only one charged handline, operated by one crew, and working from only one position within the structure. Therefore, there is no other handline or company to cover the first-arriving engine company until the supply line is established and another handline is pulled, charged, staffed, and advanced.

Even though numerous departments routinely operate in this manner, occasionally fire may rapidly spread during this stage of an incident and, in the worst-case scenario, burn through the handline as it cuts off the means of egress. In other cases, the engine operator may encounter difficulty maintaining water pressure on the interior handline during the early stages of the firefight.

Although deeply rooted in tradition, this tactic, with its associated imperfections, represents a common method practiced by firefighters nationally. Given all the drawbacks of this method, it is pure luck that there aren't more firefighter injuries or fatalities.

The single-engine evolution

During the single-engine evolution (fig. 9–2), an engine company lays a supply line for its own use. In executing this evolution, the first-arriving engine company stops at a hydrant, and after leaving a firefighter at the hydrant to connect and charge the line, lays a dry supply line into the fire scene.

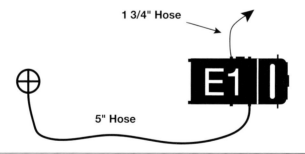

Fig. 9–2. The versatile single-engine evolution, which can be used at the start of or during an ongoing incident, involves the use of a straight (forward) lay by a single-engine company to secure a continuous supply of water. This engine company is then used in any of a number of ways, including attacking the fire with handlines, attacking the fire with exterior or interior portable monitors, attacking the fire with an elevated master stream, or pumping into a fire department connection.

Once on the scene, the remaining crew will lay and advance a handline to begin an attack on the fire while working with the water from the booster tank. In the meantime, the firefighter on the hydrant charges

9 Hose Evolutions and Fireground Realities

the supply line after the apparatus operator has connected the supply line to the engine's large inlet.

In addition to the attack-supply evolution, the single-engine evolution is widely used today on the fireground. However, it has drawbacks similar to those of the attack-supply evolution, as previously mentioned.

During execution of the single-engine evolution, time during the emergency is lost when the first-arriving engine stops to leave a firefighter and supply line at the hydrant, extending the amount of time required before the first engine crew actually enters the structure to begin a fire attack or primary search.

Even if a second engine company will not be arriving or is expected to arrive but not within a reasonable amount of time, for the safety of the firefighters on the first-arriving engine, the use of this evolution may be necessary. Moreover, it would be highly advantageous for the officer on the first-arriving engine to know if the structure involved is either residential or commercial, as this should ultimately determine which evolution is used.

For example, since fires involving residential occupancies typically require a primary search, the first-due engine should respond directly to the scene to conduct a size-up and initiate a primary search if in fact one is required. However, and although there may always be an exception, commercial structures typically involve occupants who will self-evacuate and therefore the first-due engine company officer may take the added time to stop at a hydrant on the final approach and to execute a single-engine evolution using a straight lay.

The amount of attack time and protection offered by the water carried in an engine's booster tank varies. It is therefore important for firefighters to understand that the rate of flow from a handline understandably depends on the nozzle's rated flow, whether the nozzle is fully or partially opened, and whether it is used continuously or intermittently.

For an engine with a 500-gallon booster tank, delivering 200 gpm at 170 psi engine pressure through four sections of 1¾-in. hose, the booster tank will run dry in about 2½ minutes if the nozzle operator keeps the nozzle fully open and flowing continuously. This type of use would be seen during a well-involved structure fire resulting in a defensive attack.

However, if the nozzle is half opened, providing a continuous flow of about 100 gpm, the booster tank will provide about a 5-minute supply of water before it runs dry.

The water in the booster tank will last even longer during scenarios in which the nozzle is opened for short durations and then closed during an attack. This type of interior action would be more accurately described as an attack–shut down–advance and re- attack procedure, which is perhaps the interior attack method most commonly used in the fire service.

Therefore, in objective terms, during the attack–shut down–advance and reattack procedure, the time spans when the nozzle is shut collectively extend the time before the booster tank runs dry. It is important to note that this attack process also has a favorable effect by acting to lengthen the time allowed to establish a water supply for the first engine on the scene.

Contemporary Approaches to Fire Flow Development

The dual pumping evolution

Dual pumping is a hose evolution rarely used by today's firefighters, but if executed in a different manner, it can potentially provide the added safety and effectiveness necessary to help prevent firefighter disorientation fatalities in opened or enclosed structure fires.

Generally speaking, there are two locations during an incident where this specific evolution can be carried out: at the location of the hydrant or at the scene of the fire. However, prior to discussing contemporary approaches, it is good to review the original version to better understand advantages of the modern-day methods.

9   Hose Evolutions and Fireground Realities

Dual pumping at a hydrant with 2½-in. or 3-in. supply hose

Traditionally, the dual pumping evolution takes place at a hydrant and involves a total of four engines. Once the evolution is completed, two of the apparatus serve as attack engines, working at the scene of the fire while the remaining two supply engines work in a dual connection at a single hydrant to provide the water flow (fig. 9–3).

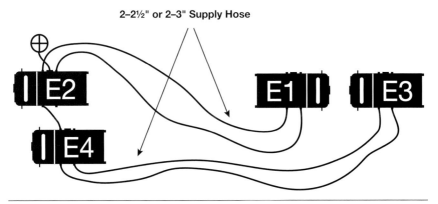

Fig. 9–3. This is the original dual pumping evolution, which uses two supply engines that share water from a single hydrant to deliver the water to two attack engines working at the scene of the fire.

When hydrants located in the immediate area of a large working structure fire are in use and more flow is still required, dual pumping is an option. In addition to providing additional flow, the evolution avoids the need to establish long lays of supply line to reach more distant hydrants.

Generally speaking, there are two ways to conduct dual pumping. The slower, traditional version uses 2½- or 3-in. supply lines, and the quicker, contemporary method uses 5-in. large diameter hose and quick-connect Storz couplings.

During the traditional dual pumping evolution, firefighters on an attack pumper attack the fire at the scene while a supply pumper connects to a hydrant and delivers water to the attack pumper through either dual 2½-in. or 3-in. supply lines. To increase the flow at the scene, another attack pumper is positioned while a fourth pumper lays additional dual 2½-in. or 3-in. supply lines from the scene to a position next to the supply pumper already pumping water from the hydrant.

Originally, to obtain additional flow from the hydrant, the second engine would line up side by side with the first to attach a hard suction hose from the first engine's steamer connection to the second engine's steamer connection. However, prior to removing the steamer cap from the engine on the hydrant, the hydrant would be closed down until a suction gauge pressure reading just above zero was obtained. The cap was then removed without having to deal with the excess residual water pressure, and the hard suction connection was completed. The hydrant was then fully opened, providing water to both supply engines. The second engine then working in a dual connection at the hydrant could pump the required flow to the second attack pumper, which was waiting at the scene for a water supply. With the introduction and use of soft suction hoses attached to gated keystone valves or piston intake relief valves (PIRVs) on steamer connections, the connection between engines was facilitated.

In the past, the traditional version of dual pumping at the hydrant successfully and effectively provided additional flow for the fire service where major flows were badly needed. However, it did require practice, space at the hydrant, and a clear understanding of the steps involved. Completing the evolution also required a total of four engines (two executing a reverse lay to the hydrant due to the need to leave the male coupling ends of the hose at the attack engines, thereby allowing a connection to the female inlets to be made on the two attack engines); the required number of sections of dual 2½-in. or 3-in. supply lines needed to complete the lay; and a certain amount of time, which would be considered by today's standards as excessive.

Dual pumping at the scene with 5-in. hose

The dual pumping at the scene with a 5-in. hose evolution, which represents a modern approach to fire flow development, takes full advantage of the high flow-carrying capability provided by 5-in. large diameter hose and low associated friction losses (figs. 9–4 and 9–5). When in use, dual pumping at the scene provides firefighters the opportunity to quickly maximize the available flow from the following:

- The two attack engines at the fire scene
- The 5-in. hose used in the supply line and 5-in. dual connection
- The hydrant
- The water main

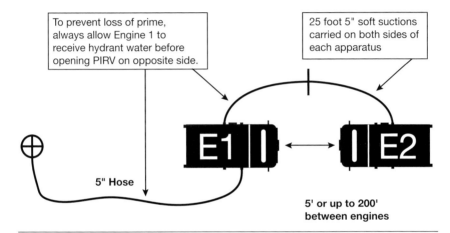

Fig. 9–4. This versatile evolution uses a steamer-inlet-to-steamer-inlet connection with 5-in. hose to enable the second arriving engine company (E2) to quickly back up the first with a handline or master stream and also receive a supply of unused hydrant water from the first engine. As pressurized hydrant water enters the pump on one side of the first engine, fire flows are developed as needed. Thereafter, unused hydrant water is obtained by opening the opposite PIRV, allowing the water to flow through the 5-in. dual connection hose and into the second attack engine's large PIRV. Flow from both engines is then maximized until a 10-psi residual reading is reached and command is notified. Note that (1) the second engine in the dual connection will be the first to reach the 10-psi residual pressure, and (2) the resulting safety by redundancy allows one engine to continue to operate should the other one experience mechanical difficulties.

Fig. 9–5. The 25-ft 5-in. soft suctions are carried on both sides of an engine to minimize friction loss and to quickly establish dual connections. As required by the incident, both engines working in a dual connection then maximize the flow by pumping for multiple handlines or master streams.

Understanding that friction loss, which is pressure loss as water flows through hose, will vary with the age of the hose, its condition, the manufacturer of the hose, and pressure gain or loss due to variations in elevation, the amount of flow ultimately delivered also depends on the distance between the fire and the hydrant or length of the hose lay.

In general, shorter lays of 5-in. hose typically provide substantial flows, while longer lays still provide good although reduced flows. Dual pumping also takes advantage of the favorable existing hydrant spacing found in communities that follow the AWWA hydrant spacing guidelines.

For example, the AWWA guidelines call for hydrants to be spaced within 1,000 ft. in residential areas. This means that the greatest distance to a hydrant in a residential area should be no more than 500 ft., that is, the midway point between two hydrants spaced 1,000 ft. apart. The spacing for commercial areas is 600 ft., which means that the greatest distance to a hydrant in a commercial area should be no more than 300 ft., the midway point between two hydrants spaced 600 ft. apart (AWWA 1999). No hydrants are provided for undeveloped open areas, as there are no structures to protect.

Firefighters have always attempted to use the hydrants that are closest to the fire as a means of minimizing the friction loss and maximizing the flow in the shortest time possible. This must also be the case with the use of 5-in. hose. When the lay is short, 5-in. large diameter hose can flow up to approximately 3,000 gpm from a hydrant with good pressure and on a suitably sized main.

Additionally, the friction loss associated with flows through 5-in. hose is advantageously low. For example, when 750 gpm, which is capable of safely and effectively extinguishing small to moderate sized residential structure fires with an additional handline dedicated to the rapid intervention team, flows through one standard 100-ft section of 5-in. hose, the result is a friction loss of 5 psi.

A major flow of 1,000 gpm through one 100-ft section of 5-in. hose is 8 psi. To obtain a flow of 1,000 gpm at a residential structure fire from a hydrant located 500 ft. away would require 40 psi. For a flow of 1,250 gpm at this same distance, 63 psi is needed; and 90 psi is required for a flow of 1,500 gpm.

To obtain a flow of 1,000 gpm at a commercial structure fire from a hydrant located 300 ft. away would require 24 psi. To obtain a flow of 1,250 gpm at this same distance, only 38 psi is required; and 54 psi is needed for a flow of 1,500 gpm.

Therefore, it is the low friction loss produced with advantageously large flows through 5-in. hose, in addition to the favorable hydrant spacing supported by appropriately sized and interconnected mains, that makes dual pumping at the scene possible. And when the closest hydrants to the scene are routinely used, the greatest flows are then quickly developed.

Therefore, conceivably, two 1,250-gpm rated engines working in a dual connection at the scene can be used to capacity (flow everything they are capable of flowing: combined 2,500 gpm) by water supplied by a single 5-in. large diameter hose attached to a strong hydrant located nearby. If not, the two engines should always use whatever flow is available to a maximum by pumping for required multiple handlines or master streams, which would result in taking the residual pressure to 10 psi—before or after a boost in water pressure is requested and received.

Maximizing the flow that represents significant firefighting efficiency is a highly desirable goal on the fireground and should be attempted by engine operators whenever possible.

On occasion, a supply engine may need to lay all of the hose in the hosebed to make a connection. However, there comes a point when due to excessive friction loss associated with greater distances, a desired flow cannot reach the attack engine, even in 5-in. hose. For these cases, the following guideline is offered for consideration.

Assuming 10 sections of 5-in. hose are carried in the hosebed, when a hose lay is equal to 10 sections or 1,000 ft. or more, an attack-supply evolution should be carried out. The attack-supply evolution, which is in actuality a two-engine relay, should be carried out in this situation because a hydrant may have a pressure of 80 psi static or less, which may not provide the larger flow required for larger structure fires at 1,000-ft distances. In these cases, the supply pumper should connect to the hydrant and pump the 5-in. line, always working within the safe operating pressure limit of 5-in. hose, which is approximately 180 psi, to the attack pumper to ensure that 1,000 gpm reaches the attack pumper. Otherwise, hose lays of from one to nine sections can be laid directly to the hydrant steamer connection.

Engine pressure for a flow of 1,000 gpm through 10 sections = 90 psi (8 psi friction loss per 100-ft section of 5-in. hose × 10 sections = 80 psi) + 10 psi for fluctuation = 90 psi in volume.

If the supply and attack engines are both rated at 1,250 gpm, engine pressure for a flow of 1,250 gpm through 10 sections = 140 psi (13 psi friction loss per 100-ft section of 5-in. hose × 10 sections = 130 psi) + 10 psi for fluctuation = 140 psi in volume.

In practice today, dual pumping at the scene can be quickly accomplished by departments having various levels of resources during the course of three different evolutions, which ultimately provide additional safety, effectiveness, and flexibility:

9 Hose Evolutions and Fireground Realities

- The single-engine evolution with dual pumping at the scene
- The attack-supply evolution with dual pumping at the scene
- The double attack-supply evolution: immediately implementing dual pumping at the scene with a three-engine response

The single-engine evolution with dual pumping at the scene

To implement the single-engine evolution with dual pumping at the scene, the first-arriving engine stops at a hydrant in the vicinity of the fire. After a firefighter has pulled and wrapped the end of the 5-in. supply line around a hydrant for anchoring purposes, the engine lays the supply line to the scene of the fire after the hydrant person has motioned or radioed for the driver to proceed into the scene. As the initial attack line is laid and charged, the kinks removed, and while a 360° size-up is being completed, the driver uncouples the 5-in. supply line from the hosebed and connects it to the gated PIRV on the large suction intake. Once the connection is made, the driver calls for water over the portable radio and the hydrant person fully opens the hydrant to charge the supply line, ensuring no kinks remain in the hose. The firefighter then joins the crew at the scene (fig. 9–6).

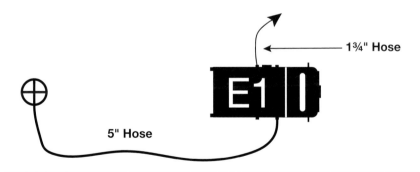

Fig. 9–6. During the single-engine evolution, Engine 1 uses a straight (forward) lay to secure its own continuous supply of water and then attacks the fire.

At any time during the course of the evolution, the next arriving engine can get into position adjacent to the first. As a part of a dual pumping SOG, the second engine will position itself within approximately 5 ft. of the first to ensure that the 25-ft. soft suction of 5-in. hose, carried on each side of each engine, will reach to allow a quick, Storz-to-Storz coupling, inlet-to-inlet dual connection to be made. If the first engine officer working in the structure has called for a backup line, the second driver charges a handline from the second engine as the company advances the line, extinguishing any fire encountered as they trace the first line to the seat of the fire and location of the first crew.

Once the 5-in. dual connection is made and after the first engine receives water from the hydrant on the opposite side of the engine, the first engine's other PIRV can be fully opened to fill the 5-in. dual connection. The PIRV on the second attack engine in the dual connection can now be opened to receive the water supply that flows through the first operating engine, making certain to remove any kinks that may have formed in the hose (fig. 9–7). Initially in this transition, however, the second driver will only partially open the PIRV on the second attack pumper because static water from the booster tank is being pumped, and fully opening the PIRV would introduce too much added pressure to the handline in operation. At the pump panel, the driver would then close the tank-to-pump valve, adjust (lower) the engine pressure, then fully open the PIRV. As in the case of the first engine driver, the second engine's driver can now charge additional needed lines to support the initial attack underway and maximize the flow as required.

9 Hose Evolutions and Fireground Realities

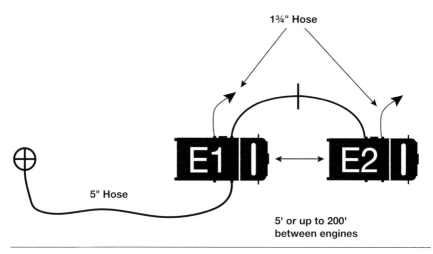

Fig. 9–7. As directed, the second or subsequently arriving engine may catch Engine 1 in a dual connection at the scene and back up the first crew by attacking the fire from the same direction.

In general, this dual connection method of operation is also used during the course of the attack-supply evolution with dual pumping at the scene and the double attack-supply evolution.

The attack-supply evolution with dual pumping at the scene (supplying with straight and reverse lays)

To implement the attack-supply evolution with dual pumping at the scene, the first-arriving engine begins by attacking the fire with either handlines or a master stream. The second engine company then lays a 5-in. supply line using either a straight (forward) or reverse lay. During the evolution and regardless of the specific lay used, the 5-in. supply line is connected directly to the hydrant's steamer connection and charged after the attack pumper's driver has connected the 5-in. hose to the large inlet (PIRV) and is ready to receive the water supply (fig. 9–8).

Preventing Firefighter Disorientation

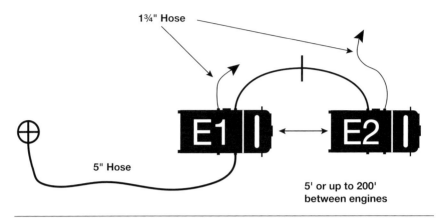

Fig. 9–8. As Engine 1 attacks the fire, Engine 2 uses a straight (forward) lay to supply Engine 1. As directed, Engine 2 backs up Engine 1 by attacking the fire from the same direction, then catches Engine 1 in a dual connection at the scene.

Immediately following completion of the lay, the second engine starts an attack as directed to support the ongoing effort and catches the first attack pumper in a dual connection at the scene. When a straight lay is used (from the hydrant to the fire) and after the 5-in. line has been connected to the first engine's large inlet (PIRV) and charged, the second engine simply positions itself close enough to the first to make the inlet-to-inlet dual connection on the opposite side of the engine. Similarly, after use of a reverse lay (from the fire to the hydrant) and after connecting the supply line to the hydrant and charging the supply line, the operator would simply drive around the block and return to the fireground to catch the attack pumper in a dual connection at the scene (fig. 9–9).

When adequate space in the area is available, it may also be possible for the supply engine to safely make a U-turn to return to the scene. After positioning on the scene and charging the attack line called for, the second company in the dual connection would then obtain water by charging the 5-in. dual connection to support the ongoing attack on the fire using either handlines or master streams.

9   Hose Evolutions and Fireground Realities

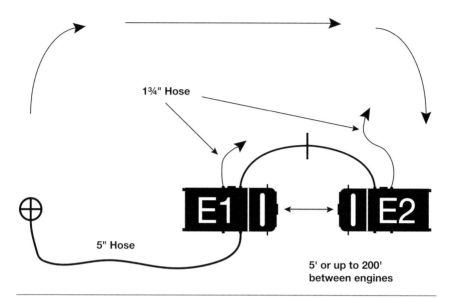

Fig. 9–9. As Engine 1 attacks the fire, Engine 2 uses a reverse lay to supply engine 1, drives around the block, then catches Engine 1 in a dual connection at the scene. As directed, Engine 2 backs up Engine 1 by attacking the fire from the same direction.

Dual pumping at the scene during the attack-supply evolution with a three-engine response

Greater flexibility and efficiency are realized when more engines are available. For example, during the early stages of an operation and when three engines are responding, the third-arriving engine can be directed to get into a dual connection with the first engine and immediately attack the fire with water from the booster tank, while the second is involved in laying the supply line (in either a straight or reverse lay) to the first engine at the scene (fig. 9–10). In all cases, the second engine driver in the dual connection must wait until after the first engine has received water before charging the 5-in. dual connection line. At any rate, within given departments, the evolutions available should be based on the actual number of staffed and equipped engines responding, and for safety and predictability, those evolutions should be a part of understood departmental standard operating hose evolution guidelines.

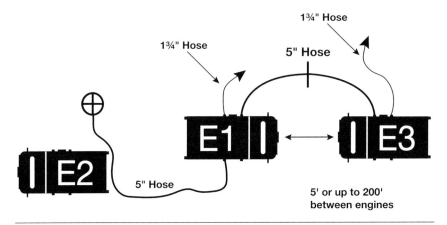

Fig. 9–10. Engine 1 attacks as Engine 2 supplies using a reverse lay. As directed, Engine 3 backs up Engine 1 by attacking the fire from the same direction then catches Engine 1 in a dual connection at the scene. Note: Engine 2 may have also used a straight (forward) lay.

The double attack-supply evolution: immediately implementing dual pumping at the scene with a three-engine response

For firefighters who respond within a jurisdiction that adheres to the AWWA guidelines, which would have reliable hydrants working from mains with adequate pressures and flows, and proper hydrant spacing, they may be able to take advantage of the double attack-supply approach to enhance safety. However, three staffed and suitably equipped engine companies stationed within reasonable distances and therefore having acceptable response times should be available to simultaneously converge on a single fire location (fig. 9–11).

Due to limited resources, some departments may not be capable of implementing this particular evolution but many are or can implement at least some beneficial aspect of it as seen in the previously described evolutions such as the single-engine evolution with dual pumping at the scene and the attack-supply evolution with dual pumping at the scene (fig. 9–12).

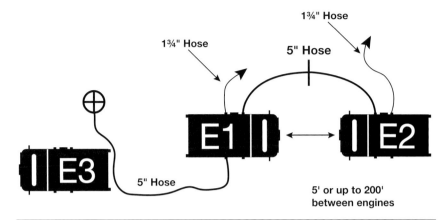

Fig. 9–11. This versatile evolution involves a coordinated attack by the first two arriving engine companies, initially using water from their booster tanks, while the third responding engine company supplies the first. The double attack-supply evolution, which incorporates dual pumping at the scene, uses a steamer-inlet (first company) to steamer-inlet (second company) connection with 5-in. hose. This dual connection enables the second company to quickly receive a supply of unused hydrant water from the first and in effect to quickly maximize the available flow. Based on numerous benefits offered, including a fast backup line to prevent flashover, greater and more rapid fire knockdown, rapid engine redundancy, and the greatest chance to achieve a successful rescue, the double attack-supply evolution should be the first-arriving officer's evolution of first consideration when resources are available.

Fig. 9–12. Although the double attack-supply evolution should be an evolution of first choice, to achieve greater flexibility, efficiency, and safety, dual pumping at the scene can also be executed after implementation of the single-engine or the attack-supply evolution. Firefighters should discuss the need for correct lengths of 5-in. hose on congested firegrounds to quickly make an inlet-to-inlet dual connection.

The double attack-supply evolution is a three-engine evolution that strives to achieve greater safety, effectiveness, and flexibility on the fireground during the earliest, most critical time of a structure fire. It does so by making the most effective use of the time, staffing, water, and equipment available at the beginning of an emergency when intervening action—such as coordinated fire attack, coordinated ventilation, and rescue—is needed the most.

In general, quicker backup and faster development of maximum flows achieved at the outset of the double attack-supply evolution equate to greater work efficiency and better outcomes. In light of this capability, use of this modern evolution should become a goal of departments with the necessary resources because it may be the determining factor in firefighter and victim survival and in minimizing the extent of property damage.

For example, consider a fire at a single-family dwelling with an opened design. Using a double attack-supply evolution, the first two arriving engines companies will immediately attack the fire with handlines in a coordinated manner and pump with a dual connection at the scene, as the third engine establishes a continuous supply of water to be used by both.

The author conducted an unpublished supply-time comparison study of 70 simulated structure fire responses to locations that were geographically well distributed in a metropolitan department (Mora 1999). This study found that, 85% of the time, the third-arriving engine company was capable of supplying the first engine on the fire scene more quickly when using the double attack-supply evolution as compared with the traditional attack-supply evolution (the blitz attack). This increase in efficiency was accompanied by other increases in firefighter safety and tactical benefits. Note that the study recorded the times required to establish a continuous water supply during the attack-supply evolution using dual 3-in. supply lines compared to utilizing the double attack-supply evolution with 5-in. large diameter hose.

Contingencies in any evolution should always be established for instances when the supply engine does not reach the scene soon enough. Therefore, as part of a dual pumping SOG, the first engine should be directed to disconnect handlines and supply the second engine should the third supply engine not be able to reach the scene in a timely manner. However, to achieve the best possible outcome, drivers on the scene or command may decide which engine lays the supply line based on the situation. For laying purposes and convenience, the location of the nearest hydrant in relation to the direction that the attack engines are pointing should determine which engine will lay the 5-in. supply line.

Historically, and with regard to the actual need for a supply line, aside from the service provided by the first-arriving engine, other responding companies may not be needed at the scene, for example, where there are smoke scares, malfunctioning alarms, false alarms, or other nuisance calls.

In other cases, the first engine knocks down the fire in the incipient stage, and there is no extension. In still other occasions, the first and second engines knock down the fire, and there is no extension, or the first engine connects to a hydrant on the scene and the supply engine does not need to supply.

Usually there is no lack of supply during the execution of the double attack-supply evolution, as the third engine generally arrives to supply both of the attack engines working in a dual connection when requested.

As part of a dual pumping SOG, flexibility on the fireground must always be provided when needed. For example, drivers should remember that a slightly longer distance may result between attack engines due to various conditions: scene congestion caused by narrow dead-end streets, narrow and dead-end roadways at apartment complexes, parked vehicles in front of the fire, or the required positioning of an aerial ladder in close proximity to the structure. In this case, drivers can use spare sections of 25-ft., 50-ft., or entire 100-ft. sections of 5-in. hose to make the dual connection (see fig. 9–12).

However, to facilitate the timely delivery of water, the supply pumper, Engine 3, may lay the supply line to the second attack engine. After the second attack engine receives the water supply, the operator on the first attack engine may now open the PIRV on the second engine and then the PIRV on the first engine to obtain water.

Additionally, due to lower friction loss when short lays are made to hydrants of adequate pressure and flow located nearby, and to provide greater flow either from handlines or master streams on a different side of a large structure, it would be safe to use up to two sections of 5-in. hose (200 ft.) between attack engines during execution of the single-engine evolution with dual pumping at the scene, the attack-supply evolution with dual pumping at the scene, or the double attack-supply evolution. In all likelihood this tactic would be utilized in a commercial district, which requires more substantial fire flows from hydrants grouped more closely together and is adequately supplied by mains of 8-in. or greater diameters. This flexibility is at times required and should be authorized as a standard emergency operating procedure during the course of a dual pumping operation.

Fire flow capability of the double attack-supply evolution

Assuming that each engine is equipped with a 500-gallon booster tank, firefighters utilizing the double attack-supply evolution are capable of quickly discharging at least 1,000 gallons of water from the first two arriving engines companies. If larger booster tanks are used, greater initial flows are achieved.

For example, if the first-arriving engine is equipped with a 500-gallon tank and the second with a 750-gallon tank, then the two-engine team initially delivers a total of 1,250 gallons of water while the third company lays a 5-in. supply line that will ultimately supply both. If both engine companies are equipped with 750-gallon tanks, they will be able to knock down 1,500 gallons worth of fire. This results in an effective two-pronged attack that will give the interior firefighters the greatest chance of achieving a successful rescue of a trapped victim while also preventing a flashover from cutting off the means of egress from either company as a supply line is laid.

When objectively examined, one can see that during the execution of the double attack-supply evolution, the volume of water discharged at the earliest stage of the incident is double the amount typically applied during the attack-supply evolution (the blitz attack) or the single-engine evolution.

One closely related fact requires mentioning. Firefighter and civilian injuries and fatalities have occurred in the past when an interior attack was underway while the second arriving engine prepared to supply or the first engine experienced mechanical or operator difficulties. The double attack-supply evolution attempts to prevent those scenarios from taking place at either opened or enclosed structure fires. Preventing the causes of life-threatening hazards as soon as possible also serves to avoid firefighter disorientation and line-of-duty deaths.

A key benefit of the double attack-supply evolution is that it allows the double engine team to maximize the available flow in the shortest amount of time during the earliest stages of an incident. In the process of maximizing the fire flow on the scene, as much of the water available from city mains is used before or, if needed, after a boost in water pressure is actually received. In contrast, during the course of a blitz attack (attack-supply evolution) with 5-in. hose, the attack engine may be effectively used to capacity, but the flow remaining in the 5-in. hose, the hydrant, and the main is not maximized and the potential safety and efficiency benefits are not realized. This outcome is reflected in the ongoing numbers of traumatic line-of-duty deaths suffered at the scene of structure fires, the number of civilian injuries and fatalities, and the property loss figures sustained nationally. Figures 9–13, 9–14, and 9–15 illustrate the fire flow capability provided by use of the double attack-supply evolution.

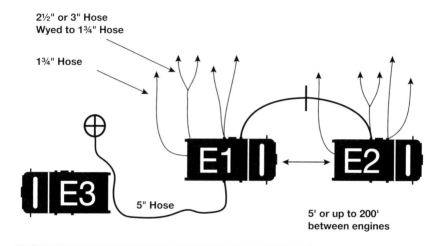

Fig. 9–13. For interior attacks and depending on hosebed configuration, multiple handlines can be quickly deployed to quickly maximize the flow using the double attack-supply evolution. In this example, 200 gpm is delivered from each 1¾-in. handline, for a flow of 1,000 gpm from each attack engine; thus, 2000 gpm from both. If six handlines from each engine have been provided, the flow from each would have been 1,200 gpm, or 2,400 gpm from both engines. In this example, and with required engine pressure, it would be good practice for command officers to estimate a flow of 200 gpm from each handline staffed and operated by an engine company. Therefore, for active fire flow determination during offensive attacks: One handline = one engine company = 200 gpm.

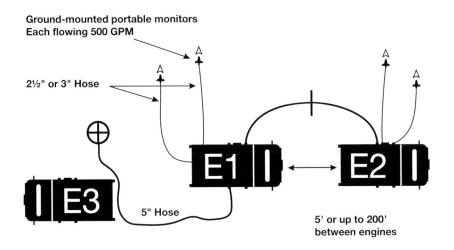

Fig. 9–14. For interior attacks at large structures and when two compact and lightweight portable monitors are immediately available on each engine, four ground-mounted monitors can be quickly deployed during the double attack-supply evolution. In this example, 500 gpm is delivered from each monitor for a flow of 1,000 gpm from each attack pumper; thus 2,000 gpm from both.

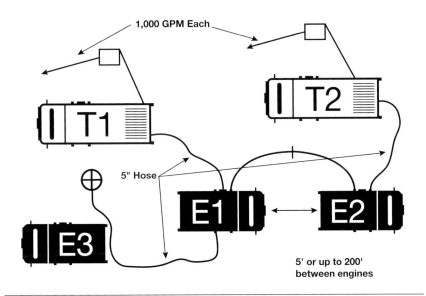

Fig. 9–15. Using the double attack-supply evolution, each elevated master stream is supplied by 5-in. hose from each attack engine, each delivering 1,000 gpm. The total combined flow from both engines in this example is 2,000 gpm.

Safe Operating Residual Pressure

In addition to guidelines that call for maintaining a 20 psi residual pressure for engine companies to prevent cavitation or a collapse of the main, engine operators may safely take the residual water pressure from the 5-in. supply line to 10 psi during the double attack-supply evolution.

According to the National Fire Protection Association (NFPA), "A minimum residual pressure of at least 20 psi (138 kPa) should be maintained at hydrants delivering the required fire flow. . . . When hydrants are well distributed and of the proper size and type (so that friction losses in the hydrant and suction line will not be excessive), it may be possible to set 10 psi (69 kPa) as the minimum pressure" (Anderson 1986).

During major firefighting incidents, engine operators must advise command when they have reached the 10 psi residual pressure mark. If, at this point, command is comfortable with progress of the firefighting effort, no increase or boost in pressure will be requested from the water utility. However, if engines reach 10 psi and the fire is not yet controlled, command may contact the dispatch office to request that the water pressure be increased or boosted in the area.

When a request for a boost in water pressure is received, the water utility will determine the location of the fire and increase the pressure. However, the amount of pressure increase operators will see on their suction gauges will only be approximately 10 psi. It is important to note that an increase in pressure of 10 psi in water mains of 8-, 12-, or 16-in. (or greater) diameter will result in a tremendous increase in volume available for use by the operating engine companies. To prevent unnecessarily high water pressure for an extended period of time, command should ensure that the water agency is notified when the boost is no longer required.

During the double attack-supply evolution and when hydrants are adjacent to the scene, the first-arriving engine may hand-stretch to the hydrant to obtain its own water supply and charge handlines or supply master streams as needed (figs. 9–16 and 9–17). The second engine then catches the first in a dual connection to charge additional handlines or supply master streams (fig. 9–18). In these cases, and if additional flow is not required, the supply engine (third engine) would be notified that a hydrant has been caught at the scene, and they will not have to lay a supply line. In the development of fire flow, it is important to keep in mind that the various volumes of water obtained from hydrants are due to the size of the main and associated water pressure. However, significant volumes of water can be quickly achieved by conducting dual pumping at the scene, and the maximum amount of water may ultimately be obtained by receiving a boost in the water pressure provided by the water utility during the course of a dual pumping evolution.

Fig. 9–16. When hydrants are adjacent to the scene, the first-arriving engine may hand-stretch to the hydrant to obtain its own water supply.

Fig. 9–17. Streams can then be immediately charged to supply handlines or master streams.

Fig. 9–18. The second engine then catches the first in a dual connection to charge additional streams as needed.

9
Hose Evolutions and Fireground Realities

Benefits of the Double Attack-Supply Evolution

The quick backup line

A major benefit of the double attack-supply evolution is the provision of a quick backup hoseline. Use of a quick backup line provides a significant degree of safety for the first-arriving engine company because it minimizes the dangerous time span during which the first crew operates alone within an involved structure.

In addition, *NFPA 1500, Standard on Fire Department Occupational Safety and Health Program*, recommends a backup hose crew, and this recommendation has been repeatedly made in fatality reports generated by the Fire Fighter Fatality Investigation and Prevention Program of the National Institute for Occupational Safety and Health (NIOSH) since 1998. It is therefore considered by leading firefighter safety advocates a highly desirable operational procedure that should be routinely practiced. However, a quick backup line has been difficult to consistently and effectively implement on the fireground for the following reason: At the outset of a structure fire and due to the limited volume of water in an engine's booster tank, it is widely considered safe practice to not allow more than one handline to be used from the first-arriving engine company that has not been supplied.

Simply stated, the water in the first booster tank belongs to the first engine company. Should another handline be deployed from the first engine and booster tank, without the first crew's knowledge, it could cause the first crew to run out of water faster than anticipated and possibly create a hazardous situation if conditions deteriorate and more water is needed. However, it is generally understood that as soon as the first engine receives a continuous supply of water, a second handline can then be deployed from that engine.

During the early stages of working structure fires, all departments are capable of establishing a backup line—eventually. The important question is: To the second, how much time does it actually take to accomplish? If backup lines are to be more effective than they have been

211

in the past, they must be deployed as quickly as possible. They may then help to prevent disorientation due to sudden flashover or a collapse of a roof or floor or due to PZVCs.

In case studies involving firefighter flashover fatalities, a review of the time required to provide a backup line has suggested that if a double attack-supply evolution had been executed during those incidents, fatalities may not have occurred. If firefighters had the opportunity to take advantage of the greater level of safety provided by a more quickly arriving backup line, even though the additional water was initially obtained from a second booster tank, in retrospect, fatalities may very well have been prevented.

In actuality, and even though measured in seconds, too much emergency time transpires after the first handline from the first engine is advanced into a structure and a backup line arrives and is advanced. During this time frame, and this is widely recognized in the fire service as the next priority, a supply line is typically laid, connected, and charged during the execution of either the single-engine or the attack-supply evolution.

However, for a backup line to effectively protect the first-arriving engine crew, provision of a second attack line and continuous water supply must develop faster. Yet in reality, and in terms of time during the early, critical stage of an incident, they develop slowly, with a backup line arriving too late in the fire.

This unfortunately represents a commonly accepted method of operation in the fire service today and a significant safety concern. What is needed to strengthen this tactical weakness is provision of a quick backup line that approaches or achieves a 200-gpm flow, deployed as soon as humanly possible.

During execution of the double attack-supply evolution, a quick backup line is available for use when the second handline, deployed from the second-arriving engine company and thus by a second crew, is placed into service at the most advantageous position in the structure as soon as the second engine company is able to reach the scene.

The combined ability of the first two arriving engine companies to work safely within the limitations of the volume of water in their booster tanks to quickly initiate a coordinated attack, while a 5-in. supply line is secured, provides enhanced efficiency and major fireground advantages, including the avoidance of firefighter disorientation.

Routine use of the double attack-supply evolution (fig. 9–19), may enable departments to avoid firefighter disorientation by specifically doing the following:

- Quickly obtaining a backup line (repeatedly recommended by NIOSH and the NFPA)
- Quickly maximizing the flow
- Obtaining rapid engine redundancy (long recommended by the NFPA)
- Quickly obtaining engine operator assistance
- Maximizing the chance of achieving a successful rescue during an opened or enclosed structure fire

Fig. 9–19. Unknown engine idiosyncrasies, mechanical problems, or human error can cause a delay in obtaining water to attack a fire. The two-operator team in the double attack-supply evolution provides driver assistance to enable flow to be established, maintained, or restored. Such flow interruptions have taken place during fatal incidents in Washington, New York, Ohio, Utah, South Carolina, and Texas.

When used during offensive or defensive operations, the double attack-supply evolution enables a coordinated attack to be initiated with twice the flow in the quickest time possible, typically delivered by twice the number of firefighters. Furthermore, during the course of an offensive attack into an opened structure, rapid protection of pathways leading back to the point of entry is established, and due to the quick cooling effect of the flow from both attack lines, firefighter disorientation and injury are prevented. These injuries are often caused by burns from exposure to flashover or backdraft or by blunt force trauma caused by partial or complete collapse of a roof.

Portable interior monitors

The ability to quickly use preconnected 1¾-in. handlines at opened structure fires, initially from two attack engines and while a supply line is secured, provides greater safety and effectiveness. However, during the course of larger enclosed structure fires, which are of greater size and contain a greater amount of and more flammable contents, the overriding tactical need is the ability to quickly maximize the flow (fig. 9–20).

Fig. 9–20. Large enclosed structure fires require large flows to be developed quickly. When multiple portable ground-mounted monitors are immediately available, firefighters may safely use them in the interior from more than one direction in an effort to cut off the spread of fire and meet or exceed high flow requirements.

The ability to quickly deploy and supply enough interior portable monitors when relatively safe to do so helps avoid life-threatening hazards that contribute to firefighter disorientation and multiple firefighter fatalities, specifically disorientation secondary to PZVCs. In simple terms, the sooner the fire stops producing heavy, blinding smoke conditions, the less time firefighters are ultimately exposed to PZVCs, which, if handline separation occurs, cause firefighters to become disoriented, exhaust their breathing air, and die.

The key to successfully accomplishing this large enclosed structure tactic is ultimately determined by the number of portable monitors carried by each arriving engine company and use of the double attack-supply evolution or dual pumping subsequent to implementation of the attack-supply evolution or the single-engine evolution. Although traditionally only one portable monitor is carried on an engine, when two compact and lightweight portable monitors (500 gpm each) are available from each of the attack engines at the scene, the engines can be quickly taken to capacity or maximized to develop a significant interior fire flow in the shortest amount of time (figs. 9–21 and 9–22). Greater flow using additional interior portable monitors (master streams of 500 gpm) from other responding attack engines can also be developed thereafter as required.

Fig. 9–21. Current portable monitors are compact and lightweight.

Preventing Firefighter Disorientation

Fig. 9–22. In addition to fixed, vehicle-mounted monitors, two portable monitors may be carried by each engine to quickly maximize engine flow.

Greatest Levels of Effectiveness

When adequate resources are available, the greatest level of effectiveness on the fireground is achieved by providing officers with the training and flexibility to use any hose evolution or combination of hose evolutions based on the situation encountered. For example, an operation at a large, working, enclosed commercial structure fire may begin with the use of the double attack-supply evolution, but based on the number of arriving second-alarm companies and location of hydrants, the single-engine or the attack-supply evolution can subsequently be executed, followed by the implementation of dual pumping at the scene if additional flow is still required. This flexible approach to obtain maximum fire flow development offers firefighters the greatest level of effectiveness and fireground safety possible.

The Hydrant Flowchart for Fire Flow Determination

To quickly, safely, and effectively attack structure fires in the early stages of an incident and therefore prevent the life-threatening hazards that lead to firefighter disorientation, a volume of water that can extinguish the potential volume of fire must be available. For example, hazards such as large enclosed commercial structures should be protected by large volumes of water used to quickly attack and extinguish fires when they occur.

In preparation for major fire incidents and in addition to the collection of other key structural and life safety information, prefire surveys give responding companies the time and opportunity to determine, in general terms, the amount of flow available from nearby hydrants for firefighting purposes. Although water mains of sufficient diameter and flow are required for commercial or residential districts, verification for anticipated tactical options that maximize the flow is always good practice.

One of the most accurate methods of fire flow determination involves recording and translating pressure readings obtained from a group of hydrants. However, another method that can quickly determine approximate flow by use of an engine's master intake gauge (suction gauge) involves the use of the 10%, 15%, 25% static-to-residual pressure drop rule. *Static pressure* is water pressure measured when still or at rest, and *residual pressure* is the remaining water pressure measured after an engine pressure has been set and water is flowing. A quick and easy way firefighters can determine approximate flow from a hydrant is by use of the static-to-residual pressure drop rule, also known as the *percentage method*.

In the field, an engine connects to and fully opens a hydrant, and the operator makes a note of the static pressure reading on the engine's suction gauge (fig. 9–23). The operator then sets a required engine pressure to flow, for example, 500 gpm from a vehicle-mounted monitor. After the correct engine pressure is set for the nozzle type used, the operator again makes note of the corresponding pressure drop, if any. This is the residual pressure reading, also taken from the suction gauge. The percentage drop determines the additional flow available from the hydrant.

Fig. 9–23. The master intake gauge or suction gauge is used to read static and residual pressures, while the master discharge gauge or pressure gauge registers the required pressure settings for handlines, master streams, supply lines, and fire department connections.

- A pressure drop of 10% or less indicates three more equivalent flows as originally flowed remain in the hydrant. Since the original flow was 500 gpm, a 10% drop means (500 × 3) = 1,500 gpm remain in the hydrant, for a total potential flow of 2000 gpm.
- A pressure drop of 11%–15% indicates two more equivalent flows as originally flowed remain in the hydrant, for a total potential flow of 1,500 gpm.
- A pressure drop of 16%–25% indicates one more equivalent flow, as originally flowed, remains in the hydrant, for a total potential flow of 1,000 gpm.
- A pressure drop greater than 25% does not indicate that no additional water remains but rather that a volume less than originally flowed remains in the hydrant.

Therefore, a high static pressure reading does not necessarily indicate a hydrant has good potential flow. It is, however, the static-to-residual pressure drop that more accurately determines the amount of water a

9

Hose Evolutions and Fireground Realities

hydrant can flow. This is independent of the additional pressure and flow, which may be received, if requested, from the local water utility during the course of a major operation.

To assist firefighters in quickly determining the potential fire flow, a hydrant flowchart is provided for fire company use (fig. 9–24). The flowchart, which is largely based on the San Antonio Fire Department Hydrant Performance Chart, offers precalculated percentage drops in 5 psi increments for static pressures ranging from a high of 125 psi to a low of 10 psi, for residual pressures ranging from 125 psi to 10 psi, and combinations in between. The number 3, 2, 1, or 0 found at the corresponding static and residual intersection box indicates the number of additional equivalent flows remaining in the hydrant above the initial flow delivered.

										STATIC														
PSI	125	120	115	110	105	100	95	90	85	80	75	70	65	60	55	50	45	40	35	30	25	20	15	10
125	3																							
120	3	3																						
115	3	3	3																					
110	2	3	3	3																				
105	1	2	3	3	3																			
100	1	1	2	3	3	3																		
95	1	1	1	2	3	3	3																	
90	0	1	1	2	2	3	3	3																
85		0	0	1	1	2	2	3	3															
80			0	1	1	1	2	3	3															
75				0	1	1	1	2	3	3														
70					0	0	1	1	2	3	3													
65						0	1	1	2	3	3													
60							0	1	1	2	3	3												
55								0	0	1	2	3	3											
50									0	1	1	3	3											
45										0	1	1	3	3										
40											0	0	1	2	3									
35												0	1	2	3									
30													0	1	2	3								
25														0	0	1	3							
20																0	1	3						
15																	0	1	3					
10																		0	0	3				

(Left side of table labeled: RESIDUAL)

Fig. 9–24. The hydrant flowchart helps to determine approximate hydrant flow (based on the San Antonio Fire Department Hydrant Performance Chart).

To learn how much additional flow is available, multiply the initial flow by the number in the box. Zero, however, does not mean no additional water can be obtained, but rather that a lesser flow, than originally delivered, remains in the hydrant.

References

Anderson, J. R. 1986. "Water Supply Requirements for Fire Protection." In *Fire Protection Handbook*, 16th ed., edited by A. E. Linville and J. L. Cote, 17–35. Quincy, MA: NFPA.

American Water Works Association (AWWA). 1999. *Distribution System Requirements for Fire Protection*, 3rd ed, 20. Denver, CO: AWWA.

10

HAZARDS OF CONSTRUCTION

The subject of building construction and associated performance during exposure to fire is a specialty requiring extensive research and analysis. The objective here is not to echo information currently available in other sources, but rather to touch on key safety measures involving the hazards of construction in order to ultimately prevent firefighter disorientation.

To safely manage working structure fires, at a minimum, firefighters should know the type of construction most prone to collapse during a fire and understand that age, lack of maintenance, and flaws in workmanship can contribute to failure. Furthermore, firefighters should also be aware of the potential for rapid collapse associated with floor-ceiling and roof assemblies as well as engineered wood I-beams.

According to the International Building Code, which has been accepted by most cities in the United States, the following are the types of building construction:

- Type I: Fire-resistive
- Type II: Noncombustible
- Type III: Ordinary
- Type IV: Heavy timber
- Type V: Wood frame

Concerning the potential for collapse of the various types of construction during a fire, the general consensus is that Type V, wood frame construction (e.g., a single-family dwelling), has the greatest potential to collapse, and Type I, fire-resistive construction (e.g., a high-rise structure), is least prone to collapse due the use of noncombustible structural members and fire protective materials. For safety, every firefighter should be aware of this potential for collapse according to construction type when conducting structural fire operations and pre-fire surveys, during strategy and tactics training, and during the initial and ongoing size-up process.

The structural strength of floor-ceiling and roof assemblies is related to their mass and method of construction. For example, lightweight wooden or steel trusses will fail rapidly if the trusses have been exposed to direct flame impingement. Consequently, firefighters should not be working above or below trusses when the trusses have been exposed to fire, because a collapse can occur at any time.

In contrast, more substantial, dimensional wooden beams have greater mass and are therefore more resistant to early failure. To assist in fireground decision making and selection of the safest tactic, firefighters who have the task of visually scanning concealed ceiling spaces with thermal imagers at structure fires should be capable of determining if trusses are in fact present in either an attic or floor-ceiling space. They should therefore be capable of quickly identifying trusses in many cases by their characteristic triangular or bow shape appearance.

Numerous types and sizes of trusses do exist, and the following photos (figs. 10–1, 10–2, 10–3, 10–4, 10–5, and 10–6) are but a few examples of truss construction widely used in residential and commercial structures. For example, during a structure fire, if trusses are encountered by firefighters scanning a ceiling space prior to advancing, the firefighters must immediately notify their company officer or command.

10 Hazards of Construction

Fig. 10–1. Lightweight wooden truss floor assemblies, which may be used in either commercial or residential construction, are structurally sound but lose strength rapidly and collapse when exposed to fire. (Courtesy of Chief Gary Bowker.)

Fig. 10–2. Lightweight steel trusses, also known as bar joists, are commonly used in the construction of large, enclosed commercial structures having vast floor areas. However, as in the case of lightweight wooden truss assemblies, if not protected by an automatic sprinkler system, they may fail at any time after being exposed to direct flame impingement. (Courtesy of Chief Gary Bowker.)

Fig. 10–3. For comparison purposes, the 2-in. × 6-in. wooden roof joists shown here provide added mass and therefore greater resistance to faster collapse under fire conditions. (Courtesy of Chief Gary Bowker.)

Fig. 10–4. The 2-in. × 8-in. floor joists shown here also have greater mass and therefore greater resistance to early floor collapse when exposed to fire. (Courtesy of Chief Gary Bowker.)

10   Hazards of Construction

Fig. 10–5. When viewed from the exterior, bowstring trusses are characterized by the roof's bowed appearance. When bowstring trusses are exposed to fire, a high potential exists for early collapse, leading to injury, entrapment, and disorientation; therefore, interior attacks should not be initiated in these types of structures. (Courtesy of Chief Gary Bowker.)

Fig. 10–6. These are scissor trusses located in the attic of a 125-year-old high school building of Type III construction. (Courtesy of Chief Gary Bowker.)

Engineered wood I-beams, also known as truss/joist I-beams (TJIs), are made of wood chips or particle board that are pressed together using combustible adhesives. They can be found in single-family and multifamily dwellings, as well as in commercial structures.

Independent fire testing of the TJIs has documented failure times of 4 minutes and 40 seconds without warning. In addition to lightweight trusses, engineered floor assemblies have also been associated with early floor collapses leading to firefighter fatalities. Firefighters must therefore be familiar with engineered wood I-beams and must understand their dangerous lack of ability to support the weight of bunkered firefighters when the I-beams are directly exposed to fire (fig. 10–7).

Fig. 10–7. Engineered wood I-beams, also referred to as truss/joists I-beams (TJIs) can collapse early when exposed during structure fires. (Courtesy of Chief Gary Bowker.)

According to NIOSH (2009),

> The engineered wood I-joist has a different cross-sectional profile than a standard solid sawn wood joist and,

in testing, burned more quickly. . . . Time-to-failure testing has been conducted by several groups, most recently Underwriters Laboratories (UL 2008); [and earlier by] Straseske and Weber (1988); [and] Weyerhaeuser [the Weyerhaeuser Company] (1986). The UL tests show that unprotected lightweight engineered floor joist (I-joists) assemblies can fail in as little as 6 minutes, and that traditional unprotected residential floor construction assemblies failed in less than 19 minutes.

Even with the best of efforts, accurately identifying use of truss construction during heavy smoke conditions may be difficult. Without this key information, though, firefighters commonly attack the fire and hope that a collapse does not occur. Developed by retired Fire Chief Gary Bowker for his community, the Blue Diamond Firefighter Hazard ID Marking System provides rapid identification of lightweight construction in commercial buildings and therefore a valuable warning of a possible early collapse (fig. 10–8).

Fig. 10–8. The Blue Diamond Firefighter Hazard ID Marking System provides rapid identification of lightweight construction in commercial buildings and therefore a warning of a possible early collapse. (Courtesy of Chief Gary Bowker.)

Other floor assemblies may also have serious defects in workmanship or may have deteriorated over time, resulting in floor collapse during a fire. Note that even though wood beams of more massive dimensions do provide a denser and stronger structural member and therefore will withstand exposure to fire for a longer period of time without failure, they too will eventually be significantly weakened by fire and ultimately collapse over a sufficiently long exposure time.

As part of their size-up procedure, firefighters must therefore try to estimate how long the fire has been burning. They must understand that the extended exposure time leading to collapse of a floor with structural members of substantial dimensions may have completely transpired shortly after firefighter arrival. If the first-arriving company officer or command has suspicions about the lack of structural integrity of any structure, all firefighters should remain outside and a defensive attack ordered.

References

National Institute for Occupational Safety and Health (NIOSH). 2009. "Preventing Deaths and Injuries of Fire Fighters Working Above Fire-Damaged Floors." http://osfm.fire.ca.gov/downloads/reports/NIOSHREPORTFireDamagedFloors.pdf.

Straseske, J., and Weber, C. 1988. "Testing Floor Systems." *Fire Command*, June: 47–48.

Weyerhaeuser Company. 1986. "Flame Penetration Ratings According to ASTM Test Method E119 Utilizing a Small Scale Furnace" (Report No. 665). Longview, WA: Weyerhaeuser Company Fire Technology Laboratory.

Underwriters Laboratories (UL). 2008. "Report on Structural Stability of Engineered Lumber in Fire Conditions" (File No. NC9140). Northbrook, IL: Underwriters Laboratories.

11

COORDINATED VENTILATION

Types of Ventilation

In the context of structural firefighting, ventilation is the act of opening a structure to allow smoke, heat, or fire to be released. As described in the next sections, there are different ways ventilation is accomplished, including horizontal, vertical, and hydraulic ventilation. Regardless of the ventilation method used, it must always be done safely for the benefit of those on the roof and within the structure.

Horizontal ventilation

A relatively safe tactic to execute, horizontal ventilation attempts to release pent-up smoke horizontally in a structure. In practice, positive pressure ventilators (PPVs) or the wind can be used to ventilate a structure horizontally by pressurizing the interior to move smoke out of the structure through a designated vent point.

Vertical ventilation

Vertical ventilation, a more dangerous tactic, is accomplished by cutting a hole in the roof to allow heat and smoke to naturally rise up and out of the structure.

Hydraulic ventilation

Hydraulic ventilation utilizes a handline's ability to entrain air from the interior to draw smoke out of a room or structure by simply opening a charged handline and flowing water outward through an open window or door.

Principles of Ventilation

Ventilation helps the entire firefighting operation by improving visibility and reducing the chance of a flashover by releasing heat trapped within the structure. The primary search can be conducted at a faster pace when even a small degree of visibility is provided by prompt ventilation of the structure.

Prompt ventilation is important during a structure fire, but by the same token, it must also be safely coordinated, especially during the course of an unprotected enclosed structure fire. Since enclosed structures seal off or exclude exterior air from entering the building during a fire, the air, and therefore oxygen, will be consumed during the fire. At lower oxygen ranges, violent backdrafts can take place if the room is hot enough and air is introduced. This can easily and unknowingly take place when a structure is ventilated without regard to all of the possible consequences and while engine and truck companies are in the structure searching for the seat of the fire or fire extension.

Fires in enclosed structures create high temperatures, and it is during these fires that violent flashovers can take place if any type of ventilation is conducted (Haessler 1986). It is therefore important for everyone involved in an enclosed structure operation to be aware of this possibility and to closely follow the ventilation instructions provided by the assessing interior engine company. This company can observe conditions by means of a thermal imager and report conditions to the incident commander and to all other companies monitoring the operation.

It is also important for firefighters to be aware of the prevailing wind speed and direction, as it may affect the operation and the possibility of air entrainment as the assessing engine company enters the building. Unlike a residential opened structure fire with possible life safety issues, it may be best not to immediately begin ventilation during the earliest stages of a commercial enclosed structure fire.

The effect on fire spread by the wind is probably the most crucial aspect of an enclosed structure fire. Therefore sector and safety officers must keep a close tab on the situation and any changes that may cause command to pull firefighters out of a structure and adjust strategy and tactics.

Vertical ventilation also comes with hazards. Vertical ventilation at many commercial enclosed structures of lightweight steel or wood truss construction may be hazardous to firefighters due to the potential for early collapse when exposed to fire impingement. For this reason, risk management should be incorporated into these operations, including working from elevated platforms or, when possible and practical, from an adjoining uninvolved structure.

Coordinating Ventilation

Probably the most important aspect associated with conducting ventilation during the course of a working fire is to ensure that ventilation is coordinated with actions carried out by the interior attack crew. More focus has been placed on controlling the air that may be entrained into the structure when the first-arriving company conducts a primary search or a search for the seat of the fire. Without keeping windows intact or controlling the door by keeping it as closed as is practical, an unfavorable wind or air flow into the structure may cause conditions to rapidly deteriorate. This aspect of interior structural firefighting must be effectively managed by every responding truck crew, or in the absence of a truck company, by first-arriving engine companies.

Simply stated, proper ventilation requires good coordination to achieve good results. If ventilation is conducted too early, additional air will increase the fire's burning rate and potentially initiate flashover or a backdraft, whereas delaying ventilation will cause the interior crew to be exposed to hot temperatures and hot products of combustion.

However, routinely incorporating the precooling procedure advocated by National Institute of Standards and Technology (NIST) and Underwriters Laboratories (UL) at residential structure fires—that is, attacking venting room and contents fires with a straight stream directed at the ceiling for 15 seconds prior to entry—should result in interior temperatures that are safer and more tenable for occupants and firefighters alike (Kerber 2013).

At any rate, the need to conduct vertical ventilation should be taken on a case-by-case basis. Moreover, the disorientation study (Mora 2003) found that when companies attempted vertical ventilation at large enclosed structure fires, it was often difficult or impossible to accomplish during the course of the operation. Furthermore, when vertical ventilation was successful, interior conditions did not appreciably improve due to the large size of the structure and heavy involved fuel load (fig. 11–1).

Fig. 11–1. Vertical ventilation was accomplishment during the Phoenix Southwest Supermarket fire of 2001. However, venting of the interior was ineffective due to factors including the heavy burning fuel load, wind speed and direction, rapid fire spread, and associated smoke development. (Courtesy of Paul Ramirez.)

References

Haessler, W. 1986. "Theory of Fire and Explosion Control." In *Fire Protection Handbook*, 16th ed., edited by A. E. Linville and J. L. Cote, 4–47. Quincy, MA: NFPA.

Kerber, S. 2013. "What Research Tells Us About the Modern Fireground." *FireRescue*. http://www.firefighternation.com/article/strategy-and-tactics/what-research-tells-us-about-modern-fireground.

Mora, W. R. 2003. "U.S. Firefighter Disorientation Study 1979–2001." http://www.sustainable-design.ie/fire/USA-San-Antonio_Firefighter-Disorientation-Study_July-2003.pdf.

12
ENCLOSED STRUCTURE TACTICS AND GUIDELINES

Operationally and routinely speaking, command and company officers should always attempt to reduce risk when managing any structure fire. In an enclosed structure fire, this specifically pertains to avoiding, as much as possible, the life-threatening hazards known to cause disorientation, serious injury, and firefighter fatality. One of the most important courses of action for reducing risk is to develop and use flexible, enclosed structure tactics and standard operating guidelines (SOGs). Furthermore, the SOGs need be unique to the available staffing and resources within each department. Within this overall effort, firefighters also attempt to minimize property loss when it can be done without needlessly risking the lives of firefighters.

During the course of enclosed structure fires, a major objective is to maintain short travel distances from the exterior to the seat of the fire. Reaching this objective may result in avoiding excessive exposure to traumatic, life-threatening hazards such as flashover, backdraft, collapse of roofs and floors, and prolonged zero-visibility conditions (PZVCs)—and thus the prevention of firefighter disorientation. The following discussion focuses on how tactics should be implemented during various types of enclosed structure fires.

Tactics for Opened Structures with a Basement

Overview

Similar to the management of other life-threatening hazards, in order to avoid disorientation and associated injuries and fatalities following the collapse of fire-weakened floors, firefighters must routinely anticipate and make every reasonable effort to avoid falling into an involved basement. Otherwise, as history has shown, the result could be the deaths of firefighters and occupants.

Basements exist in almost every community in the country, with some structures having multiple basement levels. Furthermore, certain regions of the country have a greater concentration of structures with basements than others. Because unprotected basement fires are extremely dangerous, firefighters responding within certain geographical regions have a much greater risk of falling through fire-weakened floors or of becoming disoriented or entrapped while attacking basement fires.

Additionally, every active firefighter must have a clear operational understanding that a basement fire is one of the most dangerous types of enclosed space fires. If appropriate safeguards are not taken at the outset and during the operation, the incident can suddenly result in firefighter fatalities. This awareness must be unmistakably understood, documented, and institutionalized as SOGs that safeguard against this underlying danger. Basement fires are extremely dangerous and labor intensive, necessitating adequate staffing to provide fire attack and company rehab. Guidelines should stress that aggressive interior attacks immediately on arrival should be discouraged; instead, cautious assessments to locate and attack the seat of the fire from the exterior should be used. During these operations, multiple handlines should be immediately available to safely attack the seat of the fire once located. The use of quick developing flows produced by such evolutions as the double attack-supply evolution can be highly effective and provide protection for firefighters and occupants early in the incident.

12 Enclosed Structure Tactics and Guidelines

Although information associated with the details of a structure fire can be incomplete when first received, subsequently obtained information can be valuable to the safety of the firefighters during the remainder of the firefighting operation. In other cases, key information may be provided on the initial call. Therefore, it is critical for everyone involved in the dispatch of a reported basement fire, and whenever possible thereafter, to provide responders with all key information pertaining to the specific structure involved so that the most favorable outcome can be achieved. The delivery of key information starts with assistance from civilians, followed by assistance from the emergency call taker.

Generally speaking, favorable outcomes at structure fires are achieved whenever the incident commander and responding companies receive accurate information in a timely manner. In the absence of critical information, firefighters must make worst-case scenario assumptions. For example, when single-family residential structures located in an area known to have high concentrations of basements are reported on fire, all responders must make the assumption that the structure has a basement, that the basement is involved in fire, and that the first floor is on the verge of collapsing until proven otherwise.

In simple operational terms, structural firefighting outcomes can be defined as follows:

- A favorable outcome is one in which fire is confined to one portion of a structure without experiencing firefighter fatalities.
- An unfavorable outcome is one resulting in a defensive operation with associated firefighter injuries and/or fatalities.

Possessing key incident information allows accurate tactical decisions that reduce firefighter exposure to risk. As previously emphasized, structure fires have differing degrees of danger. It is now known that opened structure fires are dangerous, and unprotected enclosed structure fires are extremely dangerous, meaning that they take the lives of firefighters at a disproportionate rate.

Avoiding the risk related to the collapse of a fire-weakened first floor above a basement requires firefighters to focus on and understand circumstances surrounding this type of danger. During these types of potentially fatal fires, two of the most important pieces of information that may favorably influence the outcome are, first, knowing if the structure has an involved basement and, second, knowing if anyone in the structure needs to be rescued. Based on this information, and therefore the way these types of basement fire scenarios are managed, will determine if fatalities can be prevented. A review of both circumstances as a basis for decision making is warranted.

Knowing if the structure has a basement and whether the basement is involved in fire

One significant problem in the fire service today stems from the fact that firefighters are equipped and trained to be action-oriented and programmed to quickly use an offensive strategy to manage structure fires without regard to the real danger associated with an opened or enclosed architectural design. The tempo of the action taken by the responding firefighters is also governed by the goal of minimizing the time to ultimately deploy attack lines into the structure. During opened structure fires, firefighters respond quickly and conduct an initial size-up quickly. When the received information is evaluated, and it is determined that the structure is relatively safe enough to enter, firefighters will initiate a fast and aggressive interior attack from the unburned side. This action involves fully bunkered firefighters quickly advancing charged handlines into the structure in an effort to locate and extinguish the fire while a primary search is simultaneously conducted.

However, the problem with opened residential structures with basements is that too frequently a fire originating in the basement is not considered or suspected. And even when it is known that the basement is on fire, conducting a quick attack of the basement fire using the safest possible means is not considered.

On one hand, the outcomes resulting in the use of an offensive strategy at opened structure fires on slab foundations are mostly favorable. On the other hand, well-documented cases have clearly shown that an offensive strategy, when used during a fire at an opened or enclosed structure with a basement, has in many cases resulted in fatalities when firefighters walked across the fire-weakened first floor and fell into heavily involved basements. Therefore, managing fires involving opened structures with basements must be more cautious and calculating (than with an opened structure with no basement) during all phases of an incident. This means that firefighters have to manage the incident in such a way that allows them to knock down the fire while simultaneously avoiding the hazards associated with the involved basement. This singular goal must be the overriding concern of every firefighter, every company, and every safety and command officer at the scene.

During extremely dangerous, deteriorated conditions, firefighters must not needlessly risk their lives to save a structure or a life that unfortunately cannot be saved. However, when lives may be saved from first-floor levels, even during involved basement fires, every effort must initially be made to maximize the chance of civilian and firefighter survival should a collapse of the first floor take place during a rescue attempt. Prior to this action, firefighters must reduce as much of the risk as possible before arriving on the scene, again by determining the following:

- If anyone is still in the structure needing rescue
- If the structure has a basement
- If the basement is involved in the fire

Knowing if anyone is still in the structure needing rescue

Because people may be in a residential structure at any time of the day or night, a primary search at a residential structure should always be a strong consideration. This is generally the case for dwellings, unless there is credible information from a reliable source that there is no one in the structure to be rescued.

On the other hand, nonresidential structures, especially those that are commercial and large in size, are typically structures in which occupants will self-evacuate when notified of a possible fire. Therefore, because of this reality and with these rescue premises in mind, the saying that directs firefighters to "search every building every time" must be modified to one advocating that if possible, every residential structure believed to be occupied should be searched, and although there may be an exception, nonresidential structures may not have a life hazard to address.

Since it is very difficult to accurately predict the need to conduct a primary search, officers must always consider this on a case-by-case basis. One repeated scenario that has tragically resulted in firefighter fatalities involved the execution of a primary search when in fact there was no one in the structure to be rescued. In the vast majority of these incidents, if notification that everyone had exited the structure reached the firefighters who were responding or who were in the structure, the firefighter fatalities may have been averted.

These types of fatalities can be prevented by training fire dispatchers about the urgency of this information—and the urgency of obtaining this information from the caller and immediately relaying the information to command and all responding companies.

Civilians must also be informed of the dangerous task that firefighters will be performing if they believe someone is still in the structure. Therefore, if reliable persons, such as occupants who have exited the structure, definitely know that all occupants are out of the structure and accounted for, this information must immediately be given to either the emergency dispatcher or any firefighter on the scene.

Every firefighter must also understand the serious ramifications concerning this information, and it must therefore be immediately relayed to command so that everyone on the fireground is made aware that a dangerous primary search will not be needed. In that case, coordinated efforts can then be made to avoid potential danger and safely locate and attack the seat of the fire.

Common causes of line-of-duty deaths during unrequired search scenarios have involved flashovers, first-floor collapses, partial or complete collapses of the roof, and exposure to rapidly spreading wind-driven fires.

In cases involving flashovers, the priority of locating and aggressively attacking the seat of the fire at the onset was not accomplished for various reasons, including staffing shortages and the desire to quickly conduct a primary search.

In cases involving wind-driven fire, simply because the wind-driven fire hazard had not yet been clearly defined in the fire service, firefighters were not aware of what constituted a dangerous condition therefore did not use tactical approaches that may have evaded the risk.

In review, the first-arriving officer must make a 360° walk-around the structure with a thermal imaging camera (TIC) whenever possible. This is done, among other reasons, to determine if occupants needing to be rescued are hanging out of upper-floor windows, and to determine the presence of a basement. If a basement is present, a determination must be made if smoke or fire can be seen venting or involving the interior of the basement. If so, everyone on the scene or en route must be immediately notified.

Look for a slope in the terrain along the foundation line or a drop-off along one side of the structure, as these are indications that the structure has a basement. A flight of steps leading to the front, side, or rear doors may also indicate the presence of a basement. Other ways to quickly determine whether a structure has a basement is by asking occupants who have evacuated the structure, referring to quickly accessible prefire plans, or drawing on familiarity with the area. Is the structure located in an area known to have basements?

As a last resort, make inspection holes into the first floor from safe exterior positions to determine if a basement exists. During this inspection, and to prevent vertical extension, it is very important that multiple charged handlines staffed by fully protected firefighters are standing by to provide a protective fog for firefighters making the cut and to quickly knock down any fire that may possibly vent through the opening made .

Once the opening or openings have been made, firefighters must aggressively insert multiple handlines with nozzles set on full-fog or semifog patterns into the openings and keep them there until effective knockdown has been achieved. Since subsequent flare-ups may also occur, charged backup handlines should remain in the immediate area after the main body of fire has been extinguished. While always assuming a worse-case scenario, good practice for basement fires calls for use of rapidly deployable, multiple-attack, and backup lines during ground- or basement-level attacks. The ability to quickly provide 8 to 10 handlines from the first two arriving engine companies should become a standard operating procedure (SOP) for opened and enclosed structure fires.

Firefighters have safely and effectively managed basement fires throughout history by knowing the risks and by understanding their own limitations. And in doing so, year in and year out, a basement fire may become a very familiar and routine job for responding firefighters.

A small contents fire originating in a basement may be easily handled and the structure promptly ventilated to restore visibility in the basement and remainder of the structure. However, firefighters must understand that certain conditions encountered at the scene of an opened structure with a basement, although on occasion giving the false impression of low risk, are in fact extremely dangerous and can be fatal. As with all other types of enclosed structures, opened structures with basements will continue to be encountered far into the future and should be anticipated on every call.

In addition to the potential for floor collapse, other hazards associated with an aggressive interior attack of a basement fire include possible exposure to PZVCs, entanglement, backdraft, and flashover. When firefighters working in an enclosed basement are exposed to a backdraft, the event is typically very violent given that ventilation is usually not possible, as it is in an opened structure due to the vent points provided by shattering multiple windows.

When a flashover occurs, fire will fill the entire room and will vent up and out of the basement staircase, making a timely rescue from above difficult and highly improbable. It will also require the evacuating firefighters to climb up an engulfed stairway, if it can be located, as it acts as a chimney for the venting of heavy smoke and fire. Each of these hazards may cause the loss of visibility and handline separation, which lead to disorientation, which can in turn lead to injuries and fatalities caused by the heat of the fire. Entanglement hazards have also taken the lives of firefighters during the course of emergency evacuations from rapidly deteriorating basement fires when a quick backup line was not available.

Furthermore, it should be clearly understood that the collapse of an entire or even a partial floor area is not needed to expose an advancing firefighter to extreme danger. In zero-visibility conditions, a small burn hole in the floor large enough to cause a firefighter's legs and torso to fall partially or completely through the hole is all that is required to expose the firefighter to extreme danger. This scenario also has the potential to take the lives of other firefighters working in the structure should a subsequent collapse of the fire-weakened floor around the hole occur during a rescue attempt.

Other cases involving opened structures with basements have demonstrated that only a minor partial collapse of a stairway leading to a lower level can cause fatalities as well. Simply stated, not knowing or overlooking whether the fire building has a basement may easily and quickly result in a multiple firefighter fatality event.

Consequently, a structure's size, age, type of occupancy, type of construction, and condition are not the only fireground factors that determine the risk associated with it. One additional and previously unrecognized factor that must be considered is the structure's enclosed design. As an integral part of common survival knowledge, every current and future firefighter must learn that one structural feature represented by an enclosed design—whether it is a feature of the entire structure, a certain floor level, or space—has major ramifications in firefighters' ability to survive during an interior attack.

As a result, a fast and aggressive interior attack initiated immediately on arrival at the scene of an enclosed structure fire may not only be ineffective in many cases, but it can also result in multiple firefighter fatalities, defensive operations, and heavy property losses. This fact is also true regardless of whether the structure is unoccupied or civilians needing to be rescued are also a part of the fire scenario. Therefore, to achieve the best possible outcome, both of these enclosed structure scenarios must be addressed.

The chronic structural firefighting risk of fatality today is directly related to the use of traditional strategy and tactics. These tactics involve a quick initial size-up process, quick decision making, and quick offensive or defensive attack philosophy for universal use on structures. It is not that the method is not effective on smaller opened structures; the problem is that it is not always effective to use in structures and spaces having an enclosed design. This is supported by documented evidence showing that the offensive strategy does not provide the desired safety and consistency during enclosed structure fires, including opened structures with basements. In 100% of the 17 cases studied, firefighter disorientation resulted in serious injury, line-of-duty deaths, or narrow escapes when fast and aggressive interior attacks were used at structures or spaces having an enclosed design (Mora 2003).

This underscores the need for every firefighter, from the commanding officer to nozzle operator, to be fully aware, on arrival, of the design of the structure involved (opened or enclosed) and of the need to use a more cautious and calculated approach during enclosed structure fires. As history has repeatedly shown, a fire originating in any structure with an involved basement will try to take the lives of overly aggressive firefighters who fail to properly calculate the risk.

Due to the uncertainty associated with the stability of the first floor, do not enter the structure until an accurate size-up has been conducted. Thereafter, use enclosed structure SOGs that employ pre-established tactics for the purposes of fire attack or rescue at basement fires. Later arriving firefighters who may not have received the initial basement warning must also be informed by early arriving firefighters or safety, company, sector, or command officers of the existence of a dangerous basement fire and of the need to use appropriate caution.

Tactics for unoccupied opened structure basement fires

Fires involving residential opened structures with basements are common. In many cases, occupants home at the time are alerted by the smell of smoke, by activating smoke alarms, or by concerned neighbors and are able to safely evacuate the structure. Occupants may exit before firefighters arrive, as firefighters arrive, or after firefighters arrive on the scene.

During a reported basement fire and when informed by occupants that everyone is out and accounted for, full attention can be placed on locating and safely attacking the seat of the fire. In the process, officers should ask the occupants where the fire is located if known. Additionally, ask the occupants for the location of the interior basement door if it is determined that it would be safe for firefighters to enter the home to access the basement for an interior attack. When it is not clear if the first floor is safe enough for firefighters to stand on, a different approach must be used, starting with a 360° size-up of the structure.

A complete 360° size-up with the assistance of a TIC, when possible, is obtained as handlines are laid and charged. During the size-up, firefighters should look for walk-out basement doors, which provide safe and easy access to the basement's interior, or any sign indicating the location of seat of the fire, such as heavy, concentrated smoke; fire showing from a basement window; or discoloration or blistering of paint or siding.

When fire is seen venting from basement windows, an aggressive exterior attack with 1¾-in. handlines should be initiated by fully bunkered firefighters. Since it is a quick developing evolution, the use of the double attack-supply evolution should be seriously considered.

When the double attack-supply evolution is not possible, dual pumping should be employed to allow fast use of a second crew and a 1¾-in. preconnected handline, a second full booster tank of water, and rapid engine redundancy.

If signs of heavy smoke or discoloration of siding are present, use equipment in that specific area that will allow for the breaching of first floor walls to provide access to cut downward through the first floor. In other cases, it may be best to make the cut just inside the back or front door while ensuring that no one enters or exits by these means.

When the wind is not a factor, it may help with visibility to initially use positive pressure ventilators (PPVs) to provide visibility by pushing heavy smoke away from the area where a cut will be made, forcing smoke through the structure and out a vent point such as a door or window on the opposite side. Once a cut through the floor is made, one handline in full fog position is to be inserted into the hole and allowed to flow in the fully open position and, if needed, until water from the booster tank is completely dumped on the fire. In this specific situation, the second handline will provide protective cover for the first crew or, at the discretion of command, used to attack the fire through a second opening cut into the floor. These handlines can be shut down at any time at the discretion of the engine officers, sector officer, or command.

After knockdown of the main body of fire, at least one handline should remain at this position to knock down any subsequent flare-ups. Since both a manageable nozzle reaction and maximum water flow are desirable during this basement operation, lower engine pressure for these handlines may initially be required. However, once knockdown from the exterior has been achieved and if determined by command that it is structurally safe for firefighters to advance handlines into the basement for final extinguishment, a higher (correct) operating pressure must be set on all handlines.

Basement fire SOGs should require that truck crews disconnect utilities and take ground ladders to the same side the floor cut is to be made. These ladders can then be laid across weakened floors to provide access to the interior basement door and basement. Truck crews should also provide ventilation by clearing out windows without making entry into the structure.

Once the volume of smoke and steam begins to subside, truck crews can then make an assessment of the structural condition of the first floor to determine if entry toward the interior basement door and downward is safe. Fire officers should be aware that although "sounding" a floor (tapping or striking to assess floor stability) has helped in the past, it has proved to be unreliable during more recent fatal floor collapses that caused firefighters to fall into involved basements.

The use of that method therefore, and when possible, should be considered on a case-by-case basis, but always in combination with visual inspection of the basement and the underside of the first floor assembly with or without the assistance of a TIC.

If after taking every possible and reasonable safety measure, a firefighter should fall through a fire-weakened first floor while using this tactic, the danger will be significantly reduced, because the fall will be made into extinguished debris rather than a working fire, which personal protective equipment is not designed to withstand.

On the other hand, if there are indications of floor weakness such as sagging or warping of the floor or serious burn-through areas, no entry is allowed and flow will be directed from the exterior until no longer required. Another option is to use 200-gpm rated piercing nozzles either downward through first floors or through basement walls from the exterior when possible. For basement fires at commercial structures, the Chicago Fire Department uses the smothering action from high-expansion foam units to reduce the risk to firefighters and extinguish extremely dangerous basement fires in their jurisdiction.

SOGs for opened or enclosed structures with basements must be drafted, approved, and added to existing departmental operating procedures and guidelines. These SOGs must then be practiced to gain familiarity and skill with each task by each involved firefighter. As in the case of any new SOG, with practice and use, the steps will become intuitive over time, and efficiency and effectiveness will increase. Most importantly, line-of-duty deaths caused by fires in basements will eventually become a thing of the past.

Always reassess the effectiveness of basement SOGs after use, and if needed, make adjustments that will improve capability and overall safety. Common fatal scenarios in opened structures with basements have included first-floor collapses and entrapment when in fact no occupants needing to be rescued were in the structure.

For safety, firefighters should not assume unreasonable risk associated with a primary search after a determination has been made that no one needing to be rescued is in a structure. Rather, excessive risk should be avoided, as the report "Rules of Engagement for Structural Firefighting: Increasing Firefighter Survival" advocates, "Do not risk your life for lives or property that cannot be saved" should prevail throughout the remainder of the incident (International Association of Fire Chiefs—Safety, Health and Survival Section [IAFC-SHS] 2012).

Firefighter fatalities have also taken place during attempts to quickly conduct a primary search for confirmed trapped occupants (fig. 12–1). These operations may involve a scenario that ultimately traps one or more firefighters and civilians in an involved basement fire following a partial or total collapse of the first floor.

Preventing Firefighter Disorientation

Fig. 12–1. In 2006 one firefighter and one civilian died during a fire that involved the basement of this small residence. Unaware of the basement fire, injuries occurred when a partial collapse of the fire-weakened first floor caused fire crews attempting a rescue to fall into the involved basement. (Courtesy of the New Jersey Division of Fire Safety.)

Tactics for occupied opened structure basement fires

As with fires involving occupied opened structures built on concrete slab foundations, fatalities of firefighters and civilians during a primary search are possible whenever a required quick attack on the fire is delayed. This unfortunate outcome can apply to fires in both opened structures and opened structures with a basement. Although basement fires are much more dangerous and challenging, access must be gained to the fire in either case and, with the hazards in mind, attacked as quickly and safely as possible. Otherwise, with the presence of adequate air, the fire will eventually transition, at times violently, to flashover stage and/or cause a collapse of the first floor.

On occasion, a charged handline for fire attack is not available during the early phase of a structure fire. During the course of a dangerous, unprotected primary search, firefighters must remember that it is risky even in full personal protective equipment to try to elude an engulfing flashover event during an opened structure fire or to gamble on whether a fire-weakened first floor will support a firefighter's weight and that of the occupant during a rescue attempt. These actions have been attempted by firefighters and have resulted in tragedies.

In actuality, rescue of occupants from involved structures is rare. Firefighters can work through long careers and never be involved in a rescue event. However, it does happen, and therefore firefighters must be effectively trained and retrained to take correct action during these scenarios. Otherwise, injuries and fatalities may result.

On a broader scale, during the course of a residential structure fire in which a primary search is required, the overall risk can be reduced if the community is knowledgeable of the danger firefighters face during a primary search. With this in mind, it should be a goal of every department to help ensure that primary searches, when initiated, are conducted for persons who are actually in the structure. Otherwise, unnecessary risks will continue to result in lost lives.

Much can be done to reduce risk prior to and during the course of working residential fires requiring a primary search. Some specific suggestions include the following:

- Train dispatchers to ask civilians if everyone has exited the home and is accounted for, then to relay the information to responding companies immediately and ensure that the message is received.
- Educate civilians to immediately contact and inform the emergency call taker or any firefighter on the scene when all occupants are out the structure and accounted for. This information must then be immediately relayed to the incident commander, who will transmit this message to all companies who are en route or on the scene.
- Train officers, apparatus operators, and firefighters to loudly ask bystanders, on arrival, if everyone is out of the structure.
- Train officers, apparatus operators, and firefighters to understand the significance of a report stating that everyone is out

of the structure and accounted for, which in turn reduces or prevents excessive firefighter exposure to danger during the course of the remainder of an ongoing incident. This requires all firefighters, particularly those who may be staging away from the scene or on the scene but have not yet entered the structure, to immediately notify command when all citizens have safely evacuated the structure. After notification, the interior companies may retreat from the structure if conditions begin to deteriorate.

With this in mind, a response to a residential opened structure with an involved basement requiring a primary search of the first floor calls for a clear understanding of the extreme danger associated with these operations. It will also require SOGs for basement fires specifically addressing rescue and nonrescue scenarios.

With the exception of actually entering the structure for search, every action required for an attack on a basement fire from the exterior with no occupants will be done during the course of an incident with occupants requiring a primary search. Objectively, two courses of action can be taken by arriving companies during this type of rescue scenario:

- Quickly enter the structure and begin a primary search, hoping that the fire-weakened floor will carry the weight.
- Quickly attack the basement fire first from the exterior prior to entering and conducting a primary search.

In managing the risk associated with occupied residential basement fires, it is important to note that history has repeatedly shown that failure to quickly attack the fire has often resulted in the loss of both firefighters and occupants as a result of partial or total floor collapses. Use of a quick developing evolution such as the double attack-supply evolution or dual pumping at the scene is highly recommended during any type of structure fire, but especially for this particular scenario.

The goal during basement fire operations, with occupants above, is to quickly knock down the basement fire from the exterior prior to entering the structure for rescue. The tactic is executed in this sequence in the hope that if a partial or complete floor collapse should occur, a fall into an extinguished basement fire or one that is significantly extinguished will not result in the fatality of the rescuers or occupants.

At the outset and as a 360° size-up is conducted, note the location of any fire showing or if there is a concentration of heavy smoke and fire showing from a certain area of the home, as this is where an attack should be initiated. In addition, try to determine the exact last known location of the occupants from any reliable person, especially those who may have safely exited the structure. If fire is showing from a basement window, immediately attack the fire as the second company provides a quick backup with a second attack line. The following photos illustrate the type of tasks required during basement fire operations. (fig. 12–2, 12–3, 12–4, and 12–5).

Fig. 12–2. Firefighters attack visible fire as ongoing size-up continues. (Courtesy of Tim Olk.)

Fig. 12–3. Firefighters breach brick veneer siding by use of a sledgehammer. (Courtesy of Tim Olk.)

Fig. 12–4. Fire becomes visible from the wall opening. (Courtesy of Tim Olk.)

Fig. 12–5. Firefighters quickly extinguish fire showing from the basement and remain in this area to knock down subsequent flare-ups. (Courtesy of Tim Olk.)

If heavy, concentrated smoke is indicating the location of the fire, breach through siding to access the fire or break out windows to create a floor cut large enough to allow an aggressive exterior attack of the basement fire from above. Use other attack lines to flow downward through other cuts, as directed by engine or sector officers. If possible, truck crews should coordinate the use of PPVs with command to try to increase visibility without introducing additional air to the fire as the cuts are made. At the discretion of the incident commander, firefighters equipped with TICs and charged handlines may then enter the structure to conduct a primary search near the location the occupant was last seen. Ground ladders should also be taken to the side from which entry will be made to help support the weight of firefighters (fig. 12–6).

Preventing Firefighter Disorientation

Fig. 12–6. Positioning a ground ladder across a fire-weakened floor can increase safety by serving as a warning of the presence of a hole in the floor and by distributing the weight of advancing firefighters over a larger number of supporting floor joists. (Courtesy of Tim Olk.)

Routinely placing ground ladders over weakened floors provides stability and an immediate means of egress up to grade level in the event of a partial or complete floor collapse. Rapid intervention teams (RITs) should be on standby as well, with additional charged handlines in the event that flare-ups should occur. As structural conditions dictate, firefighters may enter the structure for final extinguishment or to conduct a secondary search.

Each fire must be taken on a case-by-case basis, as there may be other options available to the incident commander that may improve safety or effectiveness. For example, fire from a basement can be venting heavily from the A side, yet the C side, where an occupant is known to be sleeping, may be clear of smoke and heat, allowing for a relatively safe entry by means of a window.

During residential basement fires in which an occupant is believed to be located specifically in the basement and fire is showing from basement windows, attack the fire showing to immediately extinguish or knock down the main body of fire. Enclosed structure fires are extremely hot, they hold heavy concentrations of toxic smoke, and survival is doubtful when a person is in the room of origin of any structure or enclosed space such as a basement.

In review, when fire or smoke is showing from an occupied basement, and the integrity of the first floor is undetermined, quickly attack the fire from the exterior to cool the room and stop the fire from further weakening the supporting joists and floor decking. Once conditions indicate that entry can be made, firefighters then enter to conduct a primary search of the basement.

Piercing nozzles designed for structural firefighting are available and offer another option for use during basement fires. Capable of delivering 200 gpm and greater flows, these nozzles could be carried preconnected on one 1¾-in. cross lay hoseline on each engine company that frequently responds to dangerous basement fires. During the outset of an active basement fire, and when it is unclear if entry can be made for purposes of rescue or attack, the use of multiple piercing nozzles with a flow equal to the gpm typically delivered from a 1¾-in. handline may routinely provide safer and quicker knockdown capability. There are several currently available to the fire service.

Capability of piercing nozzles

Regarding the capability of firefighting piercing nozzles for possible use at enclosed structure fires, according to Wholesale Fire & Rescue Ltd. (WFR), "These new nozzles combine the best of durability, flexibility and are light weight. They can also be driven in with a small mallet if needed. This unique nozzle (900 series) design needs to be driven in only 3.75 inches to have full penetration of the sprayer unit. Ideal for walls, and has coverage of up to 25 feet in all directions. It is powerful enough for fire in attics, basements, mobile homes, garages, vehicles, or any confined space." These piercing nozzles feature a unique safety handle that "allows a firefighter to safely hold the unit when a mallet is being used to drive the nozzle in. The handle is removable if needed. Piercing nozzles can flow about 175 gpm within a radius pattern of 25 feet" (WFR 2016).

Other piercing nozzles have a 90° bend in the piping. With use of a 10-pound sledgehammer, this particular piercing nozzle has demonstrated the ability to be driven through a floor composed of 3½-in. concrete over 22-gauge sheet metal in approximately 25 seconds. This type of floor assembly could be encountered in fire-resistive apartment high-rise structures (Profab, Inc. 2008).

Driving the nozzle through the floor would be accomplished by fully bunkered firefighters operating with the protection of charged handlines directly above the room on fire. However, the fact that firefighters would have to work directly above the room on fire with possible zero-visibility conditions and upward extension must be factored into the decision to implement this tactic.

Large Enclosed Structure Tactics: Common Operational and Safety Aspects of the Unknown–Known Guidelines

During the process of examining the 17 original disorientation case studies, Mora (2003) noted the conditions found on arrival. The smoke and fire showing on arrival and the corresponding strategy and tactics used during large enclosed structure fires were also analyzed. Although exceptions are always possible, for incidents involving large enclosed structure fires, it was determined that when nothing or a wide distribution of smoke generally contained within the structure was encountered, the location of the seat of the fire was *unknown* to firefighters; however, when a visible concentration of smoke and or fire was showing from one specific exterior area of the structure, the location of the seat of the fire was generally *known*. These observations led to the unknown–known (U–K) guidelines.

In both types of fires—known and unknown—firefighters ultimately initiated an aggressive interior attack from the unburned side, which led to unfavorable outcomes. Based on these types of conditions and lessons learned from action taken by firefighters who unsuccessfully fought large enclosed structure fires in the past, these specific circumstances may be used to guide the actions of today's firefighters and thereby increase safety and effectiveness. To use this method of initiating an attack based on knowing or not knowing where the seat of the fire is located, firefighters must first anticipate and look for these signs during the size-up process, understand what they generally mean, and then know which of two procedures to use to safely manage the incident.

Prior to discussing the steps involved in executing the U or K guideline, a review of the associated operational and safety aspects is necessary and provided at this point:

- Initial large enclosed structure size-up
- Cautious interior assessment
- Cautious interior assessment factors
- Interior attack from the initial point of entry
- Short interior attack
- Defensive attack

The initial large enclosed structure size-up

On arrival at the scene of a large enclosed structure fire, the first-arriving officer will implement the U or K guideline and, when possible, conduct a complete 360° size-up of the structure. In some cases, an accurate size-up of all four sides of a structure can be made by the officer on the initial approach. However, a detailed size-up is essential to prevent the officer from missing any critical factor that may later endanger interior firefighters.

During the time required to conduct a walk-around, the officer, with the assistance of a TIC, will note the presence of enclosed windows or doors, or secured overhead or single-swinging doors, which may serve as alternate points of entry during the operation. The officer will also look for signs of a basement; blistering paint or soot stains on any portion of exterior walls, indicating the location of the seat of the fire; cracks on a wall or separation of walls, indicating the elongation of hot unprotected steel beams; and fire showing along the edge of the upper wall and roof, both indicating a roof collapse potential and the need for a defensive attack.

However, due to the larger size of some structures, it may not be practical for the first-arriving officer to immediately conduct a 360° assessment, and therefore the task may be given to another officer. In these situations, the walk-around is delegated, and execution of the guideline may begin from a selected point of entry. As execution of the guideline is prepared for, the trained firefighter with 360° assessment responsibility observes all sides of the structure and communicates observations to command either by a face-to-face exchange or via radio. At certain structures, the firefighter conducting the 360° assessment may be unable to access the C side of the building from the A side. This may be due to other structures adjoining the B and D walls. In these situations, and when possible, consider making entry through either adjoining structure for purposes of reaching and obtaining a timely view of the C side of the involved enclosed structure. In addition to addressing the steps and importance of a 360° assessment, a more expanded explanation of the cautious interior assessment is absolutely necessary.

The cautious interior assessment

A cautious interior assessment is a carefully executed interior size-up of conditions with the use of a TIC to determine the safety of initiating an interior attack. In addition to other safety measures, such as close positioning of backup companies with charged handlines and an RIT standing by at the point of entry, it is important to stress that the assessment is conducted with the vision and advantage provided by use of a TIC. Without the TIC, the firefighters conducting the interior assessment are at risk of disorientation due to sustained loss of vision (fig. 12–7).

12   Enclosed Structure Tactics and Guidelines

Fig. 12–7. Vision through blinding smoke afforded by use of a TIC is the basis for safely conducting a cautious interior assessment. (Courtesy of Tim Olk.)

A cautious interior assessment represents a major departure from the traditional method of operation in which a quick exterior size-up is followed by a fast, blind, and aggressive interior attack. This traditional attack fails to take into account the danger of enclosed structure fires and the degree to which they are linked to line-of-duty deaths. The purpose of the controlled and methodical assessment is first to determine if an interior or an exterior attack should ultimately be made. If the assessment determines that conditions are safe enough to execute an interior attack, the assessing officer, in coordination with command, will decide how the attack will be executed.

Overall, there are three possible attack options:

- An interior attack from the original point of entry close to the seat of the fire
- A short interior attack from a different side of the structure closer to the seat of the fire
- A defensive attack a safe distance away from the structure

Before conducting an interior assessment, however, the officer making the assessment must be fully aware of assessment factors that must be collectively considered in helping to arrive at the safest and most effective attack decision.

The cautious interior assessment factors

Since the structure's enclosed design may not offer any indication from the exterior of the extent of fire within the structure, the first officer will determine whether an offensive attack may be possible or whether a defensive attack is ultimately used. Although it is not possible to discuss every imaginable factor that could be involved during the course of a cautious interior assessment, the following are key factors that the assessing officer must be aware of to reach a safe strategic decision. Once a supply line has been charged, the assessing officer uses a TIC within safe limitations and cautiously advances with a charged handline and crew, to measure the following factors:

1. The officer must consider the amount and type of content and its arrangement within the structure to determine whether or not it will hinder an advance to the seat of the fire. A haphazard arrangement will cause the crew to work harder and consume more air in the effort to negotiate hose around scattered contents, also slowing the advancement and retreat from an eventual fire attack. The fire can grow during this time span, making the volume of fire greater than originally anticipated. Due to collisions with interior objects occurring by the act of the crew advancing with the hose, stacks of contents that are lighter in weight and unstable may fall over and onto hose and firefighters. On the other hand, contents that are placed into substantial shelving or stacked and arranged in a stable and orderly fashion allow for a faster advancement and attack without the risk of having displays tip over and onto hose and firefighters in zero-visibility conditions.

2. The officer must consider the amount of smoke and heat present at the point of entry and understand that it may be different at the ceiling level. During heavy smoke conditions, the assessing officer must make use of the TIC to determine whether rollover,

which is a sign of flashover, is occurring along the ceiling. This would then require an attack to cool the interior and prevent a flashover from taking place. If interior conditions become untenable or if the volume of fire showing is determined to be too great to safely attack, the assessing company would evacuate the structure.

3. From the point of entry, the officer must consider the amount of fire showing and try to determine if the fire is consuming contents or if the contents and the structure are involved, thereby risking collapse. This will determine if an interior attack should be made and, if safe, the size and number of attack lines and total flow needed. Here are a few specific scenarios:

   a. If the size of a contents fire is manageable, that is, extinguishable with a flow from one 1¾-in. handline (200 gpm), then an attack may be initiated. In the process, the officer should always use the safety afforded by the stream's reach. For example, if a crew advances and due to the lack of additional 1¾-in. hose caused by a hang up, stops within 50 ft. of the fire, the crew should attack the fire directly from a distance or direct a solid stream pattern to the ceiling above the fire to apply as a broken stream off the ceiling and onto the fire. In the meantime and with assistance from engine companies at the point of entry, additional hose would be fed for additional reach while avoiding the formation of loops in the hose.

   b. The officer, through command, should call for an engine company to provide immediate backup for fire attack if the amount of fire showing warrants two handlines, which would result in a 400 gpm total flow from two 1¾-in. handlines, each discharging 200 gpm.

   c. If access to the seat of the fire is good, allowing effective advancement of leader lines with 3-in. feeder hose and with a manageable fire situated at a deeper position, leader lines provide another option for use by two engine companies (one company staffing each side of the hose arrangement) and sufficient backup companies.

d. If the fire requires more than approximately 400 gpm, another option involves the use of rapidly deployed portable monitors, preconnected with 3-in. hose and having a flow of at least 500 gpm, to be advanced for a heavy interior attack. Backup portable monitors should be on standby for immediate use if conditions dictate. Low-angle portable monitors are available that offer the capability of delivering 500 gpm while safely lowering the flowing tip below previous limits. This provides the ability of not only attacking fires at higher positions like ceiling spaces but also almost horizontally above walls or through open doorways if needed. Models having auto-oscillating tips for unstaffed operations are also available (figs. 12–8, 12–9, 12–10, and 12–11).

Fig. 12–8. Portable monitors with flows of 500 gpm can quickly maximize flows from attack engines.

Fig. 12–9. The low-angle capability of portable monitors is key to an effective interior attack at large enclosed structures.

Fig. 12–10. Semifog and full fog patterns provide cooling and protection for firefighters when needed.

Fig. 12–11. Fire involving contents or the structure above may be attacked unstaffed by the monitor's stream.

However, if the fire has been burning for too long and taken a significant hold of a large portion of the structure, it may be safest not to make the attack. Additionally, command, safety, and sector officers must closely consider the effects of the wind on the body of fire in relation to the position of interior firefighters and the vent point, which is typically the point of entry. Firefighters should not be in the potential flow path of fire. Consideration must also be made of the wind direction in conjunction with opened doors and windows on various sides of the structure and possible partial collapse of the roof, which may cause the interior fire to subsequently be driven downward from above by the wind.

e. If the fire is severely exposing unprotected wood beams or unprotected lightweight wooden or steel trusses, because they can fail at any time after exposure to fire, a collapse of the supporting members, the roof, and heavy roof-mounted equipment can result. A company evacuation in these dangerous scenarios should then occur without delay, a personnel accountability report provided, and a defensive attack with collapse zones prepared for.

4. Fire officers must understand that the evacuation time factor is critical during a large enclosed structure incident and may be significantly longer than the time required to advance into the interior location of the fire to attack the fire originally. The visibility may have been good and heat not a factor at the onset, and because of these favorable conditions, the time needed to reach the location of the seat of the fire may have been brief. However, the time needed to safely traverse the interior under PZVCs and while reduced to a crawl will be significantly longer and must be factored into the cautious interior assessment and the ultimate decision on the strategy and tactics to implement. Keep in mind that during deteriorating conditions, the crew needs to use survival skills that will ensure that exit of the entire crew will be safely achieved. Specifically, and since the nozzle operator and the third firefighter will not be able to see during the time they remain in the building, this will call for the officer with the TIC to guide other crew members out of the building. This evacuation will require firefighters to crawl while maintaining company integrity, negotiating around obstacles, and maintaining handline contact in PZVCs. Other evacuation factors to consider concern heat drafts generated during the fire, which may cause suspended ceiling tiles to drop, as wiring used in air handing ductwork and in electronic communications melt and hang from the ceiling. These suspended wires can cause entanglement hazards and a serious delay of either the entangled firefighter, working without vision, or of the entire crew as they assist the distressed firefighter. Contents that may have been stacked or displayed upright, even though initially appearing to be stable, may have fallen onto the hose, causing a physical handline separation from blind evacuating firefighters. Loops of hose or a pile of spaghetti hose may have also inadvertently formed within the structure, further confusing and disorienting firefighters as they attempt to exit the structure (figs. 12–12 and 12–13).

Fig. 12–12. As zero visibility develops in an enclosed structure, loops of hose or entangled hose can cause evacuating firefighters to experience confusion and disorientation. To the extent possible, formation of loops of hose, or "spaghetti," should be minimized. (Courtesy of Dennis Walus.)

Fig. 12–13. Often found in departments that have experienced disorientation, safety directional arrows help to prevent disorientation by allowing firefighters to quickly evacuate in zero visibility by feeling with gloved hands for the raised arrows and following the hose in the direction that the arrows point.

Officers must be very careful when deciding whether to make a deep interior attack, as deteriorating conditions may rapidly develop that can result in a dangerous situation at a dangerous distance from the point of entry. In this situation, either handline separation or confusing entangled handlines will result in completion of the firefighter disorientation sequence (see chap. 6).

To be considered fully prepared and qualified to safely make a deep interior attack, this scenario must be repeatedly practiced so that firefighters fully appreciate the amount of time, breathing air, and physical effort, and the degree of communication and coordination that are needed to safely exit under deteriorating interior conditions. Therefore, if during a cautious interior assessment, conditions indicate that a deep interior attack would be unsafe, do not attempt one, as the lives of firefighters would then needlessly be placed at risk to save a structure (fig. 12–14).

Fig. 12–14. Officers should not attempt a deep interior attack if the interior assessment determines conditions to be unsafe. A defensive attack with clearly established collapse zones should instead be initiated. The height of the structure's exterior wall is commonly used as a safe collapse zone distance to maintain from the structure. (Courtesy of Dennis Walus.)

The interior attack from the original point of entry

After a thermal imaging scan of the ceiling space has been conducted and a consideration of the steps involved in a cautious interior assessment made, the officer may feel comfortable with the conditions and level of safety observed. After providing command with a report of the interior conditions and with approval of command, the first engine crew will initiate an offensive strategy by initiating an interior attack. If needed, the first engine officer will request a backup line (or lines) and an interior portable monitor (or monitors). In the unlikely event that the first crew requires assistance, they will call a Mayday and be helped by the RIT, which is standing by at the point of entry. Additional engine crews will also be called upon to provide protective fire streams if necessary or to assist with leading, dragging, or carrying a distressed firefighter back to the point of entry and safety of the exterior.

The interior engine officer using a TIC may be able to guide the exterior truck crew to the specific area where a safe cut in the roof can be made for vertical ventilation. Use of the TIC eliminates the guesswork of trying to blindly determine the safest and most effective point to open the roof and greatly reduces the hazard of the vent crew falling through a fire-weakened roof.

The short interior attack

Following the steps involved in a cautious interior assessment mentioned above, if the engine officer feels uncomfortable with the interior conditions, including dangerous distances, the officer can recommend the second attack option to command to reduce the distance between the exterior and the seat of the fire by initiating an interior attack from a different side of the structure. To access the structure through an alternate attack point, if structurally sound, forcible entry or breaching by the use of cutting saws or other means may be necessary. Therefore, truck crews or technical rescue teams should be prepared to use appropriate cutting blades or other tools capable of opening windows or doors that have been covered by burglar bars, security gates, or heavy metal mesh.

In addition, plans need to be developed to ensure that firefighters receive adequate and appropriate training (and retraining) on the safe and effective use of breaching equipment. When needed, breaching tools and devices are capable of opening walls constructed of various materials used in local construction including but not limited to cinderblock, masonry, brick and mortar, wooden siding, corrugated metal siding, and reinforced wall construction (figs. 12–15, 12–16, and 12–17).

Fig. 12–15. Short interior attacks provide safety when an interior attack is attempted. Therefore, to ensure that the tactic is accomplished when needed, training on the safe and proficient use of equipment such as the concrete cutting saw shown here for breaching reinforced wall construction is a requirement for preparing for extremely dangerous large enclosed structure fires. (Courtesy of Captain Dennis Meier.)

Fig. 12–16. A power chisel is used to make a triangular cut in a reinforced wall. (Courtesy of Captain Dennis Meier.)

Fig. 12–17. Training is conducted in the use of an acetylene torch to burn through rebar on a reinforced concrete wall. (Courtesy of Captain Dennis Meier.)

The time required to complete a breach in a wall depends on the wall type. This time must be factored in when selecting a specific spot to cut through a wall in relation to the location of the seat of the fire and anticipated rate of fire spread. A quicker and more effective means to breach exterior walls involves heavy machinery operated by qualified workers. Use of various resources where ever available should be standard procedure. For example, a very large structure fire that would burn for hours or days would justify requesting and waiting for the arrival of city-owned equipment for breaching purposes.

However, readily available heavy equipment and the assistance of construction workers should also be obtained if possible. Workers and site managers know their heavy equipment's capability to punch, gouge, and safely open walls or to tear metal siding off enclosed structures and are willing to help during an emergency. Figures 12–18 and 12–19 illustrate some breaching equipment appropriate for use at large enclosed structure fires. Again, these secondary points of entry created during the fire allow attack crews access into the structure to attack the fire with the advantage of having a safe means of egress located nearby should conditions rapidly deteriorate.

Fig. 12–18. Opportunities to obtain assistance from construction workers and equipment located at nearby construction sites often exist during large enclosed structure fires and will be provided on request. Front-end loaders have the ability to lift and tear metal siding to create secondary points of entry close to the seat of the fire.

Fig. 12–19. Heavier equipment capable of safely accomplishing the breaching task should be requested or obtained at the site whenever possible. Additionally, multiple pieces of equipment may be needed to make multiple openings for added safety or simultaneous short interior attacks.

In general, when considered within the enclosed structure SOG process, a short interior attack maximizes safety and prevents firefighter disorientation by minimizing the critical travel distance from the exterior to the seat of the fire. In specific terms, this enclosed structure tactic maximizes safety by preventing firefighter exposure to flashover, backdraft, and collapse of roofs or floors. It also maximizes the efficiency of self-contained breathing apparatus regardless of rating.

By creating shorter interior travel distances, the incident commander also provides a safer overall work environment that can prevent hazardous exposure to PZVCs, which, along with the previously mentioned life-threatening hazards, may lead to firefighter disorientation, injuries, and multiple firefighter fatalities.

Additionally, when safety technology that attempts to prevent firefighter disorientation is used within the enclosed structure SOG concept and in conjunction with short interior attacks, even greater effectiveness and safety can be achieved. This also pertains to the effectiveness and safety of members of the RIT, who, when working with advantageously shorter distances, will give both rescue teams and distressed firefighters a greater chance to be a part of a successful Mayday rescue operation (figs. 12–20, 12–21, and 12–22).

12   Enclosed Structure Tactics and Guidelines

Fig. 12–20. A cutting saw is used to force open a gated door to enable a short interior attack at this enclosed structure fire involving an unoccupied auto repair shop. (Courtesy of Dennis Walus.)

Fig. 12–21. An RIT and backup engine company stand by at the point of entry in case the first engine company requires assistance. Gradually feeding a hoseline as needed for the advancing engine company minimizes the formation of loops of hose and facilitates a timely emergency evacuation should one become necessary. (Courtesy of Dennis Walus.)

Fig. 12–22. As the fire is knocked down, heavy conversion steam develops. Using TICs, ensuring company integrity, and maintaining handline contact are important in preventing the disorientation sequence from unfolding during enclosed structure fires. (Courtesy of Dennis Walus.)

The defensive attack

In the third attack option, the assessing officer and crew will immediately exit the structure and call for a defensive attack whenever the following conditions are observed during the assessment.

1. When a suspended ceiling is encountered, a firefighter will lift a ceiling tile and scan the ceiling space with a TIC. During this scan, the firefighter will look for signs of lightweight wood or steel trusses or unprotected steel and will relay this information to the engine officer. In addition, should heavy smoke and heavy fire be seen filling a substantial portion of the ceiling space, all firefighters should immediately exit, relay findings to command, and prepare for defensive operations.

2. If nothing is seen during the ceiling space scan, the crew will enter to make the interior assessment. If, during this assessment, the officer observes significant areas of fire involvement of content and/or structural involvement at, for example, a dangerous distance and consuming one half of the interior, the

crew will immediately exit by reversing their direction of travel and retracing the handline back to the point of entry. Interior fires that cannot be safely managed by the use of multiple handlines and or interior portable monitors should not be attacked, and officers should order firefighters out of the structure without delay. Command will also be advised and preparations made for defensive operations. As described here and noted in the document "Rules of Engagement for Structural Firefighting: Increasing Firefighter Survival," officers and firefighters are not required to ask for permission to exit a structure during deteriorating conditions and therefore should simply and promptly leave a structure whenever conditions warrant (IAFC-SHS 2012). An accountability sector officer and command must be advised of which companies have exited the structure and a personnel accountability report (PAR) must be requested by command.

Large Enclosed Structure Tactics: Using the U–K Method for Decision Making (Unknown–Known Location of the Fire Scenarios)

It is important for firefighters to realize that calculated action should be incorporated into large enclosed structure SOGs to attempt to avoid as much of the risk as possible before the incident occurs. For safety and task predictability, they should also be designed to be implemented in the sequence of company or resource arrival by adequately staffed, trained, and well-equipped firefighters who thoroughly understand their responsibilities. This approach helps to ensure that the first-arriving company or command officer who assumes command will not become overwhelmed by the need to make several assignments at the start of the incident while size-up, incident analysis, and operational planning are delayed at the expense of safety.

An SOG for enclosed structure fires should also follow the steps normally taken during execution of the departmental hose evolution used most often. This arrangement would then help to reduce firefighter confusion and increase efficiency and safety. Although any hose evolution may be used by any department, use of the double attack-supply evolution or one that initially incorporates dual pumping at the scene is recommended because it routinely provides the multiple benefits associated with having increased apparatus, equipment, and staffing at the earliest stage of an incident.

As in any opened or enclosed structure incident, the goal of every large enclosed structure operation is to prevent exposure to disorientation and, if possible, to safely confine the fire to the area of origin. As size-up takes place, firefighters will encounter one of two major size-up factors associated with large enclosed structure fires:

- Nothing will be showing, or light, moderate, or heavy smoke will be distributed throughout the entire structure.
- A concentration of heavy smoke and/or fire will be seen at one specific exterior area of the structure.

Focusing on these important fireground factors facilitates decision making during the early stages of a complex incident by simply matching the guideline and required action with the condition seen. In other words, each condition calls for use of a specific guideline. Firefighters will use the large enclosed structure U SOG when the location of the seat of the fire is unknown, requiring firefighters to prepare for the worst by laying and charging attack and supply lines before entering the structure. The large enclosed structure K SOG is used when the location of the seat of the fire is known. This guideline will call for firefighters to quickly attack the fire where seen, as preparations for a short interior attack if needed are made.

This tactical approach runs counter to the traditional offensive strategy, which is used when the location of the seat of the fire is unknown and involves a quick and aggressive interior attack to locate the seat of the fire or to conduct a primary search. It also calls for a quick and aggressive interior attack from the unburned side when fire is seen venting from a structure.

12 — Enclosed Structure Tactics and Guidelines

In general, an offensive strategy is relatively safe and effective during smoke-showing scenarios at opened structure fires. However, at an enclosed structure fire, the problem with using the traditional approach is that in 100% of the cases examined, its use resulted in the deaths of 23 firefighters, numerous serious injuries, and narrow escapes after a disorientation sequence fully unfolded (Mora 2003). Therefore, it should be well understood that in general, and as noted by the National Institute of Standards and Technology (NIST) and Underwriters Laboratory (UL), fire-showing scenarios involving room and contents at opened structure fires such as a moderate-size residential structure calls for considering a quick, initial 15-second straight-stream attack off the ceiling from the exterior prior to entry. For smoke-showing scenarios, use of quick and aggressive offensive strategies should be considered.

However, during "nothing or distributed smoke-showing" scenarios, enclosed structure fires call for use of cautious interior assessments, calculated decision making, and short interior attacks when or if possible. For "fire showing or fire and concentrated smoke-showing" scenarios at enclosed structure fires, a quick attack of the fire where seen is initiated, while preparations for a short interior attack are made.

Condition 1: smoke is dispersed and location of the seat of the fire is unknown

In the first large enclosed structure scenario, firefighters will see nothing, light, moderate, or heavy dispersed smoke showing from the structure. The firefighters will not see fire or any evidence of fire from the exterior of the building, making the location of the seat of the fire unknown (fig. 12–23).

Historically, these were the conditions found in the vast majority of disorientation cases studied. At the time, the fireground factors observed served as the basis for firefighters to implement established SOPs. These SOPs universally called for the use of an offensive strategy, also known as a fast and aggressive interior attack, from the unburned side to search for the seat of the fire or to conduct a primary search. However, firefighter disorientation and unfavorable outcomes resulted.

Fig. 12–23. Smoke is dispersed and the seat of the fire is unknown at this large enclosed structure fire. Conditions will therefore call for use of the large enclosed structure U guideline for unknown seat of the fire. Due to the larger size of some structures, command may delegate the task of conducting a 360° walk-around to a subsequently arriving officer. When completed, the information obtained will be provided to the officer assuming command. (Courtesy of Dennis Walus.)

Therefore, a different, more cautious approach that implements enclosed structure SOGs is taken when specific smoke conditions are seen from enclosed structure fires.

Large enclosed structure SOGs are used to prevent firefighter disorientation and fatalities. They are based on research and analysis of line-of-duty deaths occurring in enclosed structure fires. These flexible guidelines are utilized during the course of enclosed structure fires regarded as extremely dangerous. Although the goal of the guidelines is ultimately to prevent firefighter fatalities while attempting to extinguish enclosed structure fires, the means by which each department accomplishes this will depend on the level of available resources. All nationally recognized safety standards and sound firefighting principles and procedures, including use of the Incident Command System, are presumed to be used during guideline implementation (fig. 12–24).

Fig. 12–24. Implementation of the Incident Command System is a requirement for safety and effectiveness on the fireground. A quiet command post allows for efficient management of complex incidents and, most importantly, the ability to immediately hear and react to a call for Mayday.

The large enclosed structure U guideline is executed when firefighters encounter an enclosed structure measuring approximately 100-ft. × 100-ft. or greater in size, and nothing, light, moderate, or heavy but tenable smoke is showing and widely distributed on arrival; the location of the seat of the fire is unknown; and there is no life hazard present.

At this point, how will firefighters know that there is no life hazard present? Although this determination cannot be definitively made, there are strong indicators including the following:

- The call is to a well-secured place of assembly at 4 o'clock in the morning.
- The structure involves an unmaintained, vacant, and completely barred or boarded-up warehouse.
- The structure has a "for sale" sign posted and is well secured.
- The owner tells you that everyone is out of the structure and accounted for.

In addition, for firefighter safety, efficiency, and task predictability, the guideline is carried out in the sequence of company or resource arrival. This will allow the plan to be logically implemented while command focuses on the initial and continuing size-up and on the need to call a greater alarm. Additionally, departments may execute hose evolutions customarily used to deploy handlines, portable monitors, or elevated master streams, and to establish an uninterrupted and adequate supply of water.

Response and guideline steps

For illustrative and explanatory purposes only, the response used for the execution of the large enclosed structure guidelines, which may be modified to include a greater number of resources, consists of four engines, two trucks, one command officer, one safety officer, and one emergency medical unit. Four persons per apparatus and a chief's aide to assist command make up the assumed staffing level. In reality, multiple alarms with heavy subsequent response would be required.

For effective and safe management of a large enclosed structure fire, the large enclosed structure SOG is used. This SOG briefly describes the tasks required of each arriving company. Clear firefighter understanding of the responsibilities associated with guideline implementation, sound officer judgment, and strong command are essential to conducting a safe and effective operation. The following section explains the steps taken in carrying out the U guideline.

Large enclosed structure U standard operating guideline

This SOG is used for enclosed structure fires measuring approximately 100 ft. × 100 ft. or greater in size. It is implemented when smoke is dispersed on arrival, the seat of the fire is unknown, and there is no life hazard present. For safety and task predictability, it is executed in the sequence of company or resource arrival, initially using the double attack-supply evolution.

1. **First-arriving officer** identifies the enclosed structure. This officer does the following:

12   Enclosed Structure Tactics and Guidelines

    a. Identifies the building as an enclosed structure during the initial size-up.

    b. Considers the need for a primary search.

    c. Describes the structure's size, occupancy, and conditions in the initial report.

    d. Implements the enclosed structure U guideline.

    e. Determines the location of the fire from the owner or tenant when possible.

2. **First-arriving engine company** performs a cautious interior assessment.

    a. Firefighter with TIC conducts a 360° walk-around to locate enclosed windows or doors and to identify the presence of a basement.

    b. The engine company selects and parks the apparatus next to the point of entry.

    c. The engine company lays and charges a handline and prepares portable monitors for interior use.

    d. The crew positions itself next to the point of entry with a TIC.

    e. The engine company conducts a cautious interior assessment as described here:

After the required companies are in position at the point of entry and the supply line is charged, the first company conducts a cautious interior assessment, working within the limitations of the TIC. The distance the company may advance will depend on the size of the structure and conditions encountered, but in all cases kept to a minimum. When the seat of the fire is located, the officer will decide, based on the possible size of the contents fire, interior arrangement, structural integrity, and distance to the fire, to conduct one of the following:

- An interior attack will be initiated from the original point of entry.

- A short interior attack from a different side of the structure will be made.
- A defensive attack will be conducted.

Command will be advised for approval of the attack intended to be implemented. If an interior attack is initiated and assistance required, the remaining companies will, as needed, back up the attack from the same direction as coordinated by command.

3. **Second-arriving engine company** serves as backup company.
    a. The engine company parks next to and connects in a dual connection with the first engine at the scene.
    b. The engine company lays and charges a second handline and prepares portable monitors for interior use.
    c. The crew positions itself with a handline and TIC next to the point of entry.
    d. The engine company monitors radio transmissions between the first engine officer and command.
    e. The engine company reorients, rescues, or backs up the first engine company as needed.
4. **Third-arriving engine company** serves as supply, IRIT, and backup company.
    a. The engine company establishes a water supply for Engines 1 and 2 operating in a dual connection.
    b. The engine company lays and charges a third handline from Engine 1.
    c. The crew positions itself with a handline and TIC next to the point of entry.
    d. The engine company monitors radio transmissions between first engine officer and command.
5. **Fourth-arriving engine company** serves as backup and staged company.
    a. Engine and driver stage next to a nearby available hydrant as remaining crew members report to the point of entry.
    b. The engine company lays a fourth handline from Engine 2

Enclosed Structure Tactics and Guidelines

    c. The crew positions itself with a handline and TIC next to the point of entry.

    d. The engine company monitors radio transmissions between first engine officer and command.

    e. When ordered by command, the driver will obtain a water supply by straight laying a supply line to a designated side of the structure to set up as an attack engine, from which a short interior attack will be initiated.

6. **Truck Company 1** serves as the RIT at the original or secondary point of entry.

    a. The crew positions itself with tools and TIC next to the point of entry to serve as the RIT.

    b. When needed for inspection of ceiling space and using a ground ladder, firefighters will pop suspended ceiling tiles and by use of a TIC will look for and report the presence of trusses, unprotected steel, heat, smoke, or fire.

    c. The truck company will monitor radio transmissions between first engine officer and command

7. **Truck Company 2** provides utility control, coordinated ventilation, and forcible entry.

    a. The truck company closely coordinates ventilation efforts with command, controlling inlet and vent points of air to avoid causing a flashover, backdraft, or wind-driven fire.

    b. The truck company opens windows or doors as directed by command.

    c. The truck company cuts and removes security bars, plywood, sheet rock, or other enclosing material, as directed by command to provide access or means of emergency egress.

    d. The truck company creates large breaches in exterior or interior adjoining walls as directed by command. When possible and to avoid obstructions in the structure, the truck company will contact the building manager, supervisor, or an employee to learn of the most advantageous location to breach the wall.

e. The truck company monitors radio transmissions between first engine officer and command.

8. **Command**

   a. Based on reported interior conditions, command will authorize or deny requests by the assessing engine on which method of attack to use: an attack from the point of entry, a short interior attack from a different side, or a defensive attack.

   b. Command will call for the staged engine to straight lay to the required side of the structure to set up as an attack engine from which a short interior attack will be initiated. Command will also establish a dual pumping operation on that side of the building.

   c. Command will safely manage the operation by obtaining initial and periodic progress reports and by updating the incident action plan of attack.

   d. Command will call for gas and electric utility crews to disconnect service to the structure.

   e. Command will call for greater alarms and establish additional dual pumping operations.

   f. When needed, command will order all crews out of the structure early, call for PARs, and transition to a defensive attack.

   g. When needed, command will ensure that at a minimum, the following additional sectors are established: accountability, safety officer(s), interior, rehab, medical, traffic control, and public information officer. At the outset of an incident, building owners or employees can quickly provide valuable information including whether everyone has evacuated the structure, the location of the fire, and the most advantageous point to breach a wall to provide unobstructed firefighter entry during short interior attacks (fig. 12–25).

   Additionally, command is ultimately responsible for the accountability of all firefighters during all working structure fires, particularly one involving a large enclosed structure fire. For example, if six firefighters enter a building, then six firefighters must safely exit the building, and the ongoing movement of firefighters must be closely tracked by an accountability sector officer. Maintaining company integrity is critical; therefore, crews must always enter the structure together, exit together, and rehab together (fig. 12–26).

Fig. 12–25. A building manager informs a firefighter of a location that is well-suited to breach a wall for entry. (Courtesy of Dennis Walus.)

Fig. 12–26. Tracking the movement of fire crews on the fireground is an important, continuous process necessitating the establishment of an accountability sector officer.

Condition 2: smoke is concentrated in one area and the seat of the fire is known

In the second scenario and while en route or on arrival, firefighters will see smoke or both smoke and fire showing from a specific area of the structure. This was the situation encountered in three specific, large enclosed structure fires examined during the study period and three that occurred following the study period.

At the time, these observed conditions served as the basis for firefighters to implement established SOPs based on traditional strategy, tactics, and size-up factors. These traditional SOPs universally called for implementation of an offensive strategy, also known as a fast and aggressive interior attack, from the unburned side to attack the seat of the fire or to conduct a primary search. However, since the outcomes were unfavorable, a different approach should be taken for these conditions at enclosed structure fires.

During the initial size-up, whenever concentrated smoke or both smoke and fire are seen involving or venting from one area of a large enclosed structure, in keeping with the K guideline, the first-arriving officer should provide an initial report and approach the side of the structure from which smoke or fire is showing. This is typically where the seat of the fire is located and where an attack should be initiated.

However, officers must keep in mind that this condition may not always indicate the true location of the seat of the fire. There may be cases where the visible smoke is venting from the only available vent point in the structure, but the seat of the fire is actually located at a different location deeper within the structure. In these instances, a quick and deep attack should be avoided, and instead a methodical cautious interior assessment initiated to determine where the fire is located and whether a short interior attack from a different side of the structure would be safer and more effective.

Should the seat of the fire be encountered as initially suspected, utilizing the double attack-supply evolution, the first two arriving engine companies (rated at 1,000 gpm or greater) will quickly discharge both booster tanks of water (500 gallons each for a total discharge of 1,000 gallons, or 1,500 gallons when 750-gallon booster tanks are provided) to attack the fire where visible from the exterior using either 1¾-in. handlines or portable monitors, as other responding companies prepare to initiate a short interior attack.

As indicated by the double attack-supply evolution, the third-arriving engine company will lay a 5-in. supply line and assist as ordered by command. This may involve stretching a third 1¾-in. handline to establish a 600-gpm flow or advancing a portable monitor taken from one of the attack engines on the scene for developing a greater additional flow (500 additional gpm) to use in the interior if needed to achieve extinguishment.

In an effort to prevent exposure to the hazards that cause firefighters to become disoriented, engine company officers, drivers, and firefighters must clearly understand ways to quickly maximize the fire flow from their engines. Similarly, command officers must also understand the capability and limitations of all apparatus responding to the scene and therefore how to quickly maximize the fire flow from the entire response. This is facilitated whenever the appropriate SOGs are routinely used by fire departments during smaller opened structure fires.

As previously mentioned and when equipped, one way firefighters may quickly maximize the fire flow during large enclosed structure fires is by using the double attack-supply evolution and, on arrival, deploying two 500-gpm rated portable monitors from each of the two attack engines working at the scene. This would result in a combined fire flow of 2,000 gpm; that is, 1,000 gpm from each attack engine, each of which is pumping for two 500-gpm rated portable interior monitors.

Greater flows can similarly be developed by simply duplicating this action by executing another double attack-supply evolution or utilizing dual pumping at the scene. These subsequently arriving engines, if equipped, can then quickly maximize their flow by pumping for additional 500-gpm rated portable interior monitors.

The act of quickly attacking the fire as seen bypasses many of the initial steps of the U guideline that involve a careful search of the fire, thereby rapidly cooling the fire and reducing the time the fire has an opportunity to grow.

The time required to otherwise determine the type of attack to initiate is minimized as firefighters will know that they will begin by attacking the fire from the exterior as seen to reduce the fire's intensity. They will simultaneously, or as soon as additional companies arrive, make preparations to conduct a short interior attack without delay. In some cases across the country, aggressively attacking the fire as seen may be effective in immediately knocking down a major portion or all of the fire, thus preventing further fire extension.

It has been shown that large, nonresidential, enclosed structures are commonly unoccupied when firefighters arrive on the scene, as occupants typically self-evacuate. These structures are also massive and are not compartmentalized, allowing an initial exterior attack to take place while a short interior attack is organized. As well-documented history has clearly and repeatedly shown, to continue to use a traditional, uncalculated, quick and aggressive, deep interior attack from the unburned side would not provide the best possible outcome at large enclosed structure fires.

According to the disorientation study, 23 firefighters became disoriented and lost their lives in 100% of the cases examined when they used offensive strategy and tactics at enclosed structure fires (Mora 2003). In this regard, it is reasonable to assume that future outcomes at unprotected large enclosed structure fires will remain unfavorable if the same tactics continued to be used.

Large enclosed structure K standard operating guideline

This guideline is used for enclosed structures measuring approximately 100-ft. × 100-ft. or greater in size. It is implemented when smoke or fire is concentrated, the seat of the fire is known, and there is no life safety hazard. The double attack-supply evolution is initially used.

1. **The first-arriving company or command** officer assumes command and implements the enclosed structure K guideline.

    a. The officer will conduct an initial size-up and delegate a 360° assessment when the size of the structure or conditions encountered make it necessary.

    b. The officer describes the structure's size, occupancy, and conditions in the initial report.

    c. The officer in command will ensure that a life safety hazard is not involved.

    d. To quickly maximize the flow, the officer in command will ensure that the double attack-supply evolution is implemented and an attack of the visible fire is immediately initiated.

2. **The first engine company** will attack the fire using either handlines or a portable exterior or interior monitor.

3. **The second engine company** will attack the fire using either handlines or a portable exterior or interior monitor.

4. **The third engine company** will do the following:

    a. Lay a 5-in. supply line and serve as the initial rapid intervention crew.

b. When relieved by the first truck company and as directed by command, attack the fire using either handlines or a portable interior monitor.
5. **The fourth engine company** as directed by command will advance either handlines or a portable interior monitor.
6. **The first truck company** will serve as rapid intervention company at the point of entry.
7. **The second truck company** will do the following:
   a. Force entry or breach walls to enable a short interior attack to be initiated.
   b. Provide utility control.
   c. Provide coordinated ventilation.
8. **Command** will do the following:
   a. Safely manage the incident by obtaining initial and periodic progress reports and by updating the incident action plan.
   b. Call for gas and electric utility crews to disconnect service to the structure.
   c. Call for greater alarms and establish additional dual pumping operations.
   d. When needed, order all crews out of the structure early, call for PARs, and transition to a defensive attack.
   e. When needed, ensure that at a minimum, the following additional sectors are established: accountability sector, interior sector, safety officer, rehab, medical sector, traffic control, and public information officer.

Quickly maximizing the flow

During a large enclosed structure operation, firefighters should use whatever is needed to maximize the flow in order to achieve a knockdown as quickly as possible from the shortest possible distance from the exterior. When the volume of fire encountered is manageable, this may involve use of multiple and maneuverable interior handlines, each delivering maximum flows of approximately 200 gpm.

For standardization, efficiency, and safety, all engines involved are assumed to carry 1,000 ft. or 10 sections of 5-in. large diameter hose. In addition, and to minimize friction loss, maximize flow, and facilitate a rapid dual connection, two 25-ft., 5-in. soft suctions are preconnected to the engine's large intake, one on each side of the engine.

Assuming the use of 1,000-gpm rated pumpers and when working in a dual connection, two engine companies can produce a combined flow of 2,000 gpm. This will require that engines are equipped with handlines on two front cross lay compartments, a rear preconnected 1¾-in. hoseline and leader lines (two 1¾-in. hoselines) in the hosebed preconnected to 3-in. feeder hose on each of the attack pumpers. With this arrangement, each engine can potentially provide a flow of 1,000 gpm, each flowing 200 gpm from each of five 1¾-in. handlines.

Again, this would be obtained from two forward, preconnected cross lay handlines, one rear preconnected handline, and the leader lines, which consist of two additional 1¾-in. handlines.

Portable interior monitors used to attack the fire at floor level or perhaps within the overhead ceiling space can also be deployed. Assuming each engine company is equipped with one portable monitor capable of a flow of 500 gpm, both companies operating in a dual connection can deliver 1,000 gpm flow into the structure for an interior attack.

Additional monitors would be required to maximize the flow from each engine. In this regard, portable monitors from the third and fourth engine responding to the scene can be set up on the interior to cut off extension of fire and supplied by the attack pumpers at the scene to achieve a total flow of 2,000 gpm. Four 500-gpm rated monitors would then be used, two supplied by each of the attack pumpers on the scene, for a total of four monitors.

If more flow is required, the double attack-supply evolution can be duplicated by other responding second-alarm engine companies to bring the maximum total flow to 4,000 gpm by four attack pumpers on the scene.

Current portable monitors are compact, and two could be carried on each engine. However, due to the limited space on certain engines, another option in the effort to quickly maximize exterior or interior flows would be to equip all truck companies with a portable monitor for use by engine companies as needed during large enclosed structure fires.

Since use of monitors produces major flows, company, sector, and command officers should be attentive to the time monitors are allowed to flow on the fire. Shut them down when the bulk of the fire has been knocked down, and if the structure has not been compromised, allow handlines to complete extinguishment. However, as unfortunate incidents such as Pittsburgh's Ebenezer Baptist Church backdraft fire and delayed steeple collapse illustrate, if there is any doubt as to the integrity of the structure, establish safe collapse zones and accomplish final extinguishment from elevated positions on the exterior (NIOSH 2006).

Runoff in Residential versus Large Enclosed Commercial Structure Fires

Although water damage should always be avoided during the course of fire operations, runoff of water and water damage from the attack at large enclosed structure fires will occur and should be expected. Additionally, the runoff may be more than normally seen during the course of a residential structure fire.

Nonetheless, remember that firefighters are dealing with an extremely dangerous large enclosed structure fire that requires a larger fire flow. Therefore, since high efficiency of water conversion may not always occur, a larger amount of runoff will typically be the norm. Thus, firefighters should realize that heavy water damage caused by the firefight may take place to prevent an unfavorable outcome, which in the past involved defensive operations, total property loss, and multiple firefighter fatalities.

According to the disorientation study, "In 12 cases (71%) the strategy changed from an offensive to a defensive operation but only after injury or fatalities had occurred" (Mora 2003). This clearly shows that initiating fast, blind, and deep interior attacks from the unburned side, with a minimum number of handlines that produce flows acceptable for opened residential fires, may not control larger volumes of fire and should be avoided.

However, with knowledge gained and a sound understanding of the real risks, there will be some occasions when a small contents fire in a deep location of a large enclosed structure can be attacked from the unburned side following a cautious interior assessment and accomplished safely without the need to breach a hole along an exterior wall. The success of this strategy requires thorough initial and continuous training in the use of enclosed structure tactics.

Primary Search Requirements in Residential versus Nonresidential Large Enclosed Structures

In general, without the advantage of having received any specific information from dependable persons on the scene, primary searches are required at residential structures regardless of whether they have an opened or enclosed design. This is simply because people live, eat, and sleep in residences on a 24-hour basis. However, a primary search of a nonresidential large enclosed structure, requiring firefighters to make a right-hand or a left-hand turn search pattern will generally not be needed, as occupants typically exit the structure on their own and on many occasions before firefighter arrival.

In other cases, the fire will occur after business hours and the structure involved will be secured. In still other cases, the structure will either be vacant or abandoned and secured. Nonetheless, when fire or heavy smoke is visible from a nonresidential large enclosed structure, every reasonable effort must be made to verify that all occupants have exited the structure. Although it is always advantageous to have easy rules that indicate when a primary search is definitively needed, due to many possible factors, the need to conduct a primary search in a residential or a large enclosed commercial structure must be made on a case-by-case basis at the time of the fire.

References

International Association of Fire Chiefs—Safety, Health and Survival Section (IAFC-SHS). 2012. "Rules of Engagement for Structural Firefighting: Increasing Firefighter Survival." Retrieved August 11, 2013, from http://websites.firecompanies.com/IAFCsafety/files/2013/10/Rules_of_Engagement_short_v10_2.12.pdf.

Mora, W. R. 2003. "U.S. Firefighter Disorientation Study 1979–2001." Retrieved from http://www.sustainable-design.ie/fire/USA-San-Antonio_Firefighter-Disorientation-Study_July-2003.pdf.

National Institute for Occupational Safety and Health (NIOSH). 2006. "Career Battalion Chief and Master Fire Fighter Die and Twenty-Nine Career Fire Fighters are Injured during a Five Alarm Church Fire—Pennsylvania." http://www.cdc.gov/niosh/fire/reports/face200417.html.

Profab, Inc. 2016. "Fyrestick vs. Chicago Concrete and Steel Floor." http://fyrestick.com/fyrestick-piercing-nozzle-common-uses.

WFR Wholesale Fire & Rescue Ltd. N.D. "Piercing Nozzles: 900 Series Piercing Nozzle." http://www.wfrfire.com/NOZZLES/SPECIAL/pierce_900.asp.

13

FIREFIGHTER DISORIENTATION CASE REVIEWS: CONSIDERING ENCLOSED STRUCTURE TACTICS

Operations at large enclosed structure fires are complex, highly challenging, and extremely dangerous. For these reasons, the greatest reduction of risk must be made before a fire ever occurs. Without this preparation, the incident commander and responding companies may easily become overwhelmed by the situation at the expense of safety.

Due to the low frequency of large enclosed structure fires at the local level, it may be unclear what needs to be considered during the initial and ongoing size-up and tactically executed on the fireground. When this occurs, firefighters naturally tend to fall back on traditional strategy and tactics suitable for opened structures that always worked for them in the past. However, and as history has shown, firefighters seem to execute those tactics, only more aggressively. In light of the disorientation risk potential, that would be a miscalculation if attempted during an enclosed structure fire.

Time has allowed for careful analysis of case studies and, more importantly, the development of approaches that will hopefully help minimize risk and maximize safety and effectiveness. For the professional development and safety of firefighters, this chapter looks at five case study summaries of the National Institute for Occupational Safety and Health (NIOSH) Fire Fighter Fatality Investigation and Prevention Program, and the Phase II Report of the City of Charleston Post Incident Assessment and Review Team. The prevention of additional losses in other departments could not take place without the information shared in these important documents. The fire service is therefore grateful for both investigative teams and for the departments and communities involved as in all preceding case studies reviewed in this book.

The disorienting fires described in this chapter took the lives of 19 firefighters because offensive, opened-structure tactics were used at large unprotected enclosed structure fires, some of which were wind-driven. For each incident, a brief analysis of the basis for the strategy decision reached is provided, followed by a discussion of how enclosed structure fire tactics and other operational information presented in this book may have been used during those and similar fires to help achieve a better outcome.

During these reviews, the assumption is made that in addition to nationally recognized safety standards, the firefighting response includes the use of the Incident Command System, the 2-in/2-out rule, adequate staffing, adequate apparatus, adequate tools and equipment, adequate fire flow, and completed training on the understanding and safe implementation of large enclosed structure fire standard operating guidelines (SOGs); however, since none of the departments involved had the full benefit of all of these items, they all operated at a significant disadvantage.

The Worcester Incident

In December 1999, Worcester, Massachusetts, firefighters responded to a six-story cold storage warehouse fire (fig. 13–1). On arrival, light to moderate smoke was showing, and wind was from the south at 7.5 mph and was not a factor. Based on the initial size-up, an offensive strategy was used to locate and attack the seat of the fire on the second floor and to ventilate the structure. During this time frame, it was reported that

13 Firefighter Disorientation Case Reviews: Considering Enclosed Structure Tactics

homeless people might be in the structure. The order was then given for firefighters to conduct a primary search. However, even though a truck company was able to conduct vertical ventilation, and an interior attack was underway, it was ineffective, and as operations progressed, the smoke conditions within the structure began to deteriorate and over a 2-minute period, moderate smoke conditions worsened to zero visibility for the remainder of the operation.

Fig. 13–1. Unknown at the time, the Worcester Cold Storage Warehouse was an extremely dangerous large unprotected enclosed structure. During the worst-case scenario that unfolded during this incident, it became impossible for firefighters to promptly and effectively ventilate and safely evacuate the structure. (Courtesy of Scott LaPrade.)

These prolonged zero visibility conditions (PZVCs) resulted in the disorientation and fatality of six seasoned firefighters, who were not in contact with a handline to serve as a lifeline out of the smoke-filled floors. It also resulted in serious injuries and close calls for many other firefighters involved in this dangerous operation. The incident commander halted the firefighter rescue operation due to the possibility of losing additional firefighters and transitioned to a defensive operation and ultimately the recovery of fallen firefighters, which lasted for days. The homeless people who had been in the structure had escaped prior to the arrival of the responding firefighters (NIOSH 2000).

The Worcester incident analysis

The firefighters most likely based their decision to initiate a fast and aggressive interior attack on the light smoke conditions showing on arrival, which suggested that the fire was at an early stage, and on the perceived degree of danger associated with a vacant commercial structure (fig. 13–2). Without knowledge of the extreme danger associated with unprotected structures having an enclosed design and the ineffectiveness of an offensive strategy, the firefighters became participants in a firefighter disorientation sequence. Although one thermal imaging camera (TIC) was available, the heat of the fire caused the camera to "white out," precluding its use. Had it been available earlier in the incident, and if "white out" had not occurred, the TIC might have helped to some degree.

The Worcester incident with enclosed structure tactics

With an awareness of the hazards associated with unprotected enclosed structure fires, a department is capable of reducing the risk before the fire in many ways. On arrival, firefighters encountered light to moderate smoke showing from the roof; however, this sign alone did not accurately reveal how long the fire had been burning or where the seat of the fire was located. Therefore, the U (unknown) guideline would have applied here.

13 Firefighter Disorientation Case Reviews: Considering Enclosed Structure Tactics

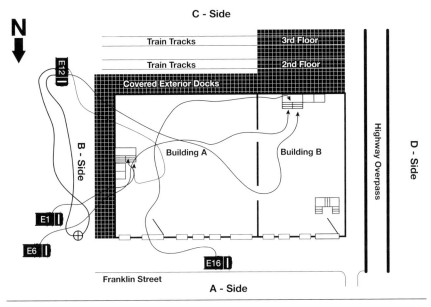

Fig. 13–2. This diagram indicates the action of Worcester firefighters as they quickly advanced and charged numerous 2½-in. and 1¾-in. attack lines to the second and third floors of the warehouse. Numerous other apparatus and personnel were also on the scene but are not depicted. (Adapted from NIOSH F99-47; see NIOSH 2000).

"When the first alarm was struck at 18:15, the fire had been in progress for about 30 to 90 minutes" (NIOSH 2000). Therefore, and as in the case of other large enclosed structure fires, the light smoke showing from the warehouse indicated that there was in fact a significant fire underway somewhere in the structure. Should a similar fire occur in any community today, a different approach could be used by departments staffed, trained, and equipped to do so. Those not properly equipped or staffed should not attempt these tactics, because ineffective and unsafe attempts may result in firefighter injury.

A 360° size-up would have been conducted, but due to the large size of the structure, it would have been delegated to and conducted by a trained officer, firefighter, or battalion chief's aide. The information obtained from the walk-around would have then been reported to command. Unlike the signs associated with an opened structure fire, it has been determined that light, moderate, or heavy smoke showing from an enclosed structure tells firefighters that they will potentially be dealing with an extremely dangerous situation in which they may be exposed to multiple life-threatening hazards including, but not limited to, flashover, backdraft, collapse of roofs or floors, PZVCs, and rapid fire spread associated from induced ventilation or from a wind-driven fire. Any of these hazards can cause disorientation, which can lead to the injuries and/or fatalities of multiple firefighters.

During a fire such as the Worcester incident, in which the location of the seat of the fire was initially unknown, firefighters would implement the U guideline, previously incorporated into existing high-rise SOGs. However, the point of entry would be based at each stairwell doorway leading to each floor area. As a cautious interior assessment is conducted on the second floor, where the fire was in fact located, the assessing officer would make an attack decision based on the fireground factors that were actually present on that day and during the incident, including:

1. Wind speed and direction
2. Structural design (opened or enclosed)
3. Life safety presence
4. Size of the structure
5. Occupancy type
6. Construction and special insulating features of the structure
7. Location, size, and extension of the fire
8. Maze-like conditions
9. Entanglement hazards
10. Holes in floors
11. Absence of interior doorknobs
12. Absence of natural or artificial light with potential for PZVCs
13. Interior doors closed flush with interior walls
14. Absence of an operable automatic sprinkler and standpipe system
15. Inherent danger associated with large unprotected enclosed structure fires

13 Firefighter Disorientation Case Reviews: Considering Enclosed Structure Tactics

In this case, the attack decision would include evacuating the structure, allowing the fire to burn, and implementing a defensive attack.

With safety officers, accountability officers, and rapid intervention teams (RITs) in place, if a report is received that homeless persons may be in the structure, integral companies of firefighters with TICs and portable radios and in constant contact with lifelines should scan floor areas while in the proximity of stairwell doorways. The search should be done as safely as possible in a reasonable attempt to determine whether occupants are or are not in the building. Again, in general, nonresidential, large, enclosed vacant warehouses usually do not need a primary search; as shown by past cases, when persons are present and not overcome by carbon monoxide poisoning, they typically evacuate on their own and usually before the arrival of firefighters.

The need to conduct a primary search of any structure should always be considered on a case-by-case basis. After this point during the incident, command should conduct a personnel accountability report (PAR) and ensure that companies remain on the exterior at the scene and, as needed, call for additional companies. When fire vents from the structure, a defensive attack should be initiated that establishes safe collapse zones, protects exposures, and maximizes the flow. In this way, firefighters are not needlessly exposed to extreme danger to save a type of structure now known to repeatedly take the lives of firefighters.

The Memphis Incident

In 1999 Memphis, Tennessee, firefighters responded to large unprotected enclosed commercial structure fire. The report also indicated that someone may have still been in the structure. On arrival, wind was not a factor and light smoke was visible from the roof. Following an initial size-up, a truck crew was ordered to check the roof, while the front doors were forced open to gain entrance.

On the interior, two officers observed smoke light enough to easily see the illuminated exit sign at the rear of the store. They attempted to locate the seat of the fire as they walked to the rear storeroom. There, in the B-C corner, the officers heard sounds of a fire behind a door leading to an office. One officer investigated and found a working fire upon opening the door.

Due to the fire's powerful suction for air, the officer was unable to reclose the door and immediately called for a handline. The fire at this point had involved the office and was extending to the ceiling space above the office area and the ceiling space above and toward the front floor space. Crews immediately advanced two 1¾-in. handlines to the rear for an attack on the fire. As is common practice throughout the fire service, a truck company checking for fire extension pulled suspended ceiling tile in the vicinity of the fire. Unfortunately, this caused a violent backdraft and partial collapse of the roof and associated building materials. The engine company officer immediately called for an emergency evacuation, but due to the developing heavy smoke and debris, which fell onto both the hose and firefighters, they became disoriented in the structure.

An RIT was pressed into service, and members were able to effect the rescues of disoriented firefighters. However, conditions deteriorated so rapidly that reentry was not possible without first knocking down the intense fire that had heavily involved the structure. Two firefighters ultimately died of their injuries despite the risk taken by fellow firefighters to help them escape (NIOSH 2004b). Prior to firefighter arrival, the person suspected of being in the structure at the time of the fire had actually stolen from the business and set fire to the office before leaving.

The Memphis incident analysis

The firefighters most likely based their decision to initiate a fast and aggressive interior attack on the light smoke conditions observed, which suggested that a fire at an early stage was underway; on the perceived degree of danger associated with a commercial structure; and on the need to conduct a primary search for a possible trapped occupant (fig. 13–3).

Without the knowledge of the extreme risk associated with unprotected large enclosed structures and ineffectiveness of an offensive strategy in those types of structures, the firefighters became unknowing participants in a firefighter disorientation sequence.

13 Firefighter Disorientation Case Reviews: Considering Enclosed Structure Tactics

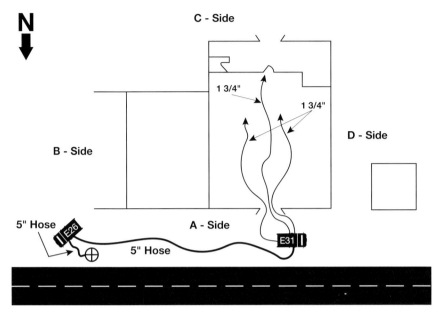

Fig. 13–3. This diagram shows the actions of Memphis firefighters as they quickly advanced handlines deep into the structure from the unburned side. The deepest handline was advanced 300 ft, with others advanced 200 and 150 ft. (NIOSH F2003-18; see NIOSH 2004b.)

The Memphis incident with enclosed structure tactics

During an incident such as the Memphis incident and since the location of the seat of the fire was unknown and with a possible trapped occupant involved, sound officer judgment based on appropriate training will come into play. In these types of enclosed structure fires the objective is to achieve a primary search and an attack on the fire while avoiding the risk associated with the enclosed structure.

In the case of an occupied opened structure with a working basement fire, that would be accomplished by first attacking the basement fire to prevent the fall of firefighters and occupants into an involved basement. In this case, the first-arriving officer would implement the U guideline, ensure a 360° size-up is conducted, and select a point of entry. Assuming the front door is selected and using the double attack-supply evolution, a handline would be laid and charged from the first two arriving engine companies while the third lays a 5-in. supply line. In this scenario, all responding companies would execute their portion of the guideline based on their sequence of arrival while the first engine company and truck company enters to begin an assessment and a search for occupants.

At the front of the store, the truck crew will lift suspended ceiling tiles which were present and scan the ceiling space. The presence of lightweight steel trusses would have been seen and it is possible that the fire at the B-C corner of the building may have also been visible. The first engine officer and the subsequently arriving command officer would have been immediately advised of this information.

Concerning a primary search and although a person may become trapped in any type of structure, there were clues in this incident which indicated the likelihood that no one was in the structure. The first was involvement of a nonresidential enclosed commercial structure in which occupants typically exit on their own accord. The second was the secured front door, which required forcible entry. The third was the light smoke on the interior, which would not have caused rapid incapacitation of occupants.

When determined that no occupants are in the structure, transmit an all clear on the primary search. When the fire is located and determined to involve an enclosed room, provide the highest degree of safety for the interior firefighters first before attacking the fire. In this case, immediately force open the rear double doors and exit while positioning companies at the rear of the structure to implement a short interior attack from the C side of the building.

Do not open any door that is immediately adjacent to a possible working fire without first advancing charged handlines or interior portable monitors capable of accomplishing knockdown of the fire as it may become impossible to close the door afterwards.

On the C side of the structure, companies are then to carry out steps in the guideline including establishment of backup companies, RIT, accountability, and safety. Command should reposition to the C side to obtain a clear view of the action taking place and of the conditions from this vantage point as ongoing size-up continues. At the rear secondary point of entry, truck crews should likewise pull ceiling tiles if any, to check for extension in the concealed space while all companies are on the exterior and while anticipating a backdraft.

Another option could include the use of a 1¾-in. handline in the straight stream setting to pop the ceiling tiles with a water stream from a safe position to reduce exposure to effects from a possible backdraft. Should a backdraft occur, and if safe, consider attacking the fire using interior portable monitors; otherwise transition to a defensive operation.

In this way firefighters are not exposed to multiple life-threatening hazards including backdrafts or a collapse of the roof while located in a deep interior position. Exposure to PZVCs and heat of the fire can also be avoided and the best possible outcome is achieved. If in such a fire a backdraft does not occur, companies can attack the room fire and any extension from the doors providing the shortest distance from the exterior using multiple handlines and or if necessary portable unstaffed interior monitors to ensure a quick and overpowering knockdown. Firefighters should then immediately exit the building while the fire is knocked down. Allow a few minutes to pass while watching for the development of conversion steam, which is of course the sign of fire extinguishment or conversely for continued production of dark smoke, indicating the structure or contents are still burning. When satisfied that a knockdown has been achieved, command may direct companies to shut down the monitors and if structurally sound, reenter with handlines to complete final extinguishment of any remaining fire or hot spots.

The Carthage Incident

On February 2004, in a mutual aid response, a career firefighter was unable to exit the interior of a large unprotected commercial structure fire involving a restaurant and lounge (fig. 13–4). With winds blowing at approximately 21–26 mph from the south-southwest, heavy smoke was showing at the unoccupied large enclosed structure. As firefighters initiated an aggressive interior attack through the main entrance on the A side, they encountered light smoke conditions.

Preventing Firefighter Disorientation

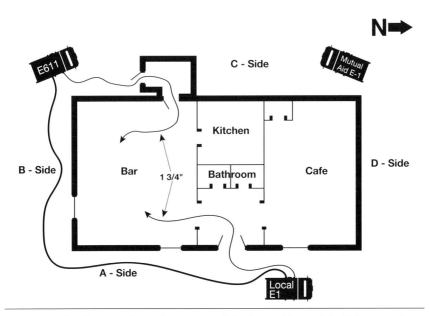

Fig. 13–4. This diagram shows the actions of Carthage, Missouri, firefighters as they advanced handlines from the A and C sides of the building. (NIOSH F2004-10; see NIOSH 2004a.)

They then advanced toward the right (D) side of the structure, where a café was located, in search for the seat of the fire. With a career firefighter equipped with a helmet-mounted TIC in the lead, the three-person crew then moved to the B side of the structure, which housed a bar, bandstand, and dance floor. As entry was made into this area, heavy black smoke causing zero visibility conditions along with worsening heat conditions were encountered.

The deteriorating conditions were caused by fire that was rapidly spreading through lightweight wooden trusses in the concealed ceiling space. During this effort, the two firefighters working the handline evacuated the structure—the nozzle operator left the building due to losing the seal on his mask, and the other firefighter retreated after mistakenly thinking the career firefighter had exited the structure earlier. This left the career firefighter alone in zero visibility with deteriorating fire conditions.

At this point, the incident commander noticed the roof above the south portion of the structure sagging, which was above the position

13 Firefighter Disorientation Case Reviews: Considering Enclosed Structure Tactics

of the interior firefighter. Command then immediately called for an emergency evacuation of the structure due to the possibility of partial collapse, which did occur soon after, near the south (B) side of the structure. Firefighters simultaneously making an interior attack from the C side safely exited. However, a PAR revealed that one firefighter was missing.

Repeated attempts by RITs were unable to locate the missing firefighter. After approximately one hour of defensive operations, the firefighter was located along the interior A wall, 25 ft from the main entrance. The disoriented firefighter was found entangled in a chair, although not to a point that would have prevented his escape from the building. When located, his face mask was not covering his face (NIOSH 2004a).

The Carthage incident analysis

Firefighters most likely based their decision to initiate a fast and aggressive interior attack on the light smoke conditions showing on the interior and on the perceived degree of danger associated with the unoccupied commercial structure. This decision may have also been based on the visual capability provided by a helmet-mounted TIC used during the incident. Without knowledge of the extreme risk associated with wind-driven, unprotected, large enclosed structure fires, one firefighter became an unknowing participant in a firefighter disorientation sequence.

The Carthage incident with enclosed structure tactics

Because heavy smoke was showing on the exterior and light smoke on the interior of the structure and there was no indication of the location of the seat of the fire, the U guideline would have been implemented. This guideline calls for companies to assume the worst-case scenario, set up the arriving companies, and methodically execute the SOG that incorporates a 360° walk-around and a cautious interior assessment before deciding on the safest and most effective course of action. However, firefighters must always consider the wind speed and direction and how they will affect operations.

With an awareness of the hazards associated with wind-driven fires involving unprotected large enclosed structures, every department is capable of reducing risk before the fire. Wind-driven fires involve all types of structures, from high-rise buildings and single-story residences to large enclosed structures. Should a similar fire occur today, a different approach would be used by departments trained, staffed, and equipped to do so. Those not properly trained, equipped, or staffed should not attempt these tactics, as ineffective and unsafe attempts may result in injury to firefighters.

A 360° size-up would have been conducted by the first-arriving officer or, due to the size of the structure, by a later-arriving trained officer, firefighter, or battalion chief's aide. The information obtained from the walk-around, including the location of a window on the B side and a door on the C side, would have then been reported to command.

The most important size-up factor concerning a fire such as the Carthage incident is the wind speed and direction in relation to the position of the fire within the structure and any possible vent points. This is critical information, upon which tactical decisions will be based. In this regard, firefighters should not advance into a structure from a door that could serve as a vent point for fire without first controlling the inlet, because the fire could become pressurized on another side of the building at some point during the fire.

In this type of scenario, a traditional attack from the unburned side could result in conditions that could immediately disorient and trap firefighters in the dangerous flow path of fire. In this case, risk reduction begins at the dispatch office responsible for monitoring and transmitting current and future wind speeds and directions to firefighters on every reported structure fire. An additional notification from the dispatch office of the presence of lightweight wooden trusses in the ceiling space of a structure that is large, unprotected, and enclosed would also be very beneficial and possibly alter the strategy. Additionally, all responding companies would then follow steps outlined in a wind-driven fire action plan.

13 Firefighter Disorientation Case Reviews: Considering Enclosed Structure Tactics

Common wind-driven fire scenarios involve vented fires on an exterior side of a structure that are pressurized by a 10 mph or greater wind speed. There was, however, a twist in the way the wind-driven fire in this case exposed the interior firefighter. The fire in the concealed ceiling space above was subsequently forced downward in the area of the south or B side of the structure onto the firefighter, following the partial collapse of the roof.

The vent point for the pressurized fire was provided by the doors at the main entrance on the A side while the inlet for pressurized air was created by the opening in the collapsed roof and an open door at the rear C side. With this in mind, firefighters must be aware that a fire that may vent may not be visible on the initial 360° size-up. However, it is equally critical for firefighters to understand that fire can vent at a subsequent time during the course of the fire, and therefore they can become exposed to a hazardous delayed wind-driven fire.

During the incident, the action plan would have stipulated that the doors on the A side remain closed, eliminating it as a vent point, while companies lay and advance handlines with the wind at their backs from the rear C side. While proper "door control" is exercised upon entry, ceiling would be pulled, the ceiling space scanned with a TIC, and the hidden fire attacked with multiple handlines by use of 200-gpm rated piercing nozzles or with portable interior monitors.

However, since there was no life safety issue or need to conduct a primary search, firefighters would have focused greater attention on minimizing risk during the fast spreading fire. Because the roof was supported by a lightweight wood truss, which is prone to early collapse when exposed to fire and due to fire rapidly spreading within the concealed ceiling space, an order for a defensive attack would have significantly reduced the risk associated with exposure to PZVCs, collapse, wind-driven fire, firefighter disorientation, and entanglement. When complex fires that seem to require a need for action to be taken quickly such as the Carthage fire are encountered, every officer must always keep in mind that when life safety is not an issue, the lives of firefighters should never be needlessly put at risk to save a structure.

The Phoenix Incident

The 2001 Phoenix, Arizona, Southwest Supermarket fire is pictured in fig. 13–5. The following is from the NIOSH Fire Fighter Fatality Investigation and Prevention Program report F2001-13:

> On March 14, 2001, a 40-year-old male career fire fighter/paramedic died from carbon monoxide poisoning and thermal burns after running out of air and becoming disoriented while fighting a supermarket fire. Four other fire fighters were injured, one critically, while fighting the fire or performing search and rescue for the victim. The fire started near a dumpster on the exterior of the structure and extended through openings in the loading dock area into the storage area, and then into the main shopping area of the supermarket. The fire progressed to five alarms and involved more than 100 personnel. Fire fighters removed the victim from the structure and transported him to a local hospital where he was pronounced dead. (NIOSH 2002)

Fig. 13–5. The 2001 Phoenix Southwest Supermarket fire. (Courtesy of Paul Ramirez.)

13 Firefighter Disorientation Case Reviews: Considering Enclosed Structure Tactics

For orientation purposes, the following series of fireground diagrams and photographs provide interior and exterior views of the structure (figs. 13–6, 13–7, and 13–8).

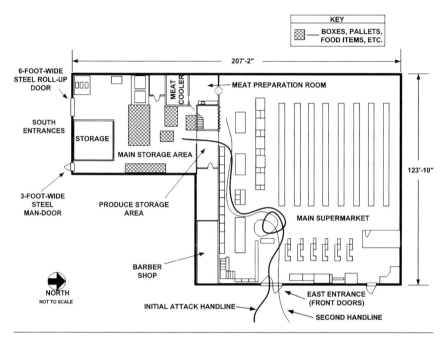

Fig. 13–6. This diagram indicates the location of the south and east (front) entrances, which served as the points of entry where the initial attacks were made. It also specifies the location of the storage room and remainder of the supermarket floor space. (NIOSH F2001-13; see NIOSH 2002.)

Fig. 13–7. Plywood that partially covered the windows on the A side of the structure were removed and windows broken to provide ventilation and greater visibility in the structure. (Courtesy of Paul Ramirez.)

Fig. 13–8. This photo is of the main storage room and contents negotiated by rescue crews during the difficult removal of the disoriented firefighters. Consider the fire flow required to cool and prevent the lightweight steel trusses from collapsing during the rescue operation. (NIOSH F2001-13; see NIOSH 2002.)

13 Firefighter Disorientation Case Reviews: Considering Enclosed Structure Tactics

On arrival, wind was from the south (B side) at 15 mph with heavy smoke showing on the B side of the structure. Occupants of the grocery store and adjacent stores were out when an inspection was made by an early arriving firefighter. The first-arriving engine company attacked the dumpster fire, which was difficult because of arcing power lines. They also made an attack of the rear storage room fire, which initially was not thought to have been involved.

At the same time, Engine 14 was ordered to check the interior of the grocery store for conditions. Light smoke was showing on the interior of the sales floor area. However, an inspection of the rear storeroom revealed a working fire. Handlines were then immediately advanced to the rear storeroom, where companies played streams on the fire to prevent extension into the main sales area.

During this time frame, visibility dropped to zero, and at that point one firefighter informed his officer that he was low on air and needed a replacement cylinder. The officer then instructed his crew to get on the handline and follow him out of the structure. When the crew exited, it was learned that two firefighters were not with them. A Mayday was transmitted by one of the disoriented firefighters at the same time that a change to a defensive attack was being planned. RITs were pressed into service to locate and help the firefighters out of the building, while roof ventilation took place and windows at the front of the store were broken to provide ventilation and greater visibility in the structure.

During this effort, fire traveled rapidly through the structure, preventing egress from the front, or A side, of the structure. Multiple handlines were used to aggressively attack the fire in the rear storeroom, prevent the roof from collapsing, and allow one firefighter to locate an interior crew by moving toward the sound of their voices, and he was able to obtain their assistance in evacuating.

During this time, RITs worked with great difficulty to bring the other disoriented and disabled firefighters out of the structure through the doors on the B side. The incident then transitioned to a defensive attack, during which 12 calls for Mayday were transmitted at this extremely dangerous, wind-driven, large enclosed structure fire (fig. 13–9). (See NIOSH 2002.)

Fig. 13–9. This photo indicates the B (south) side of the grocery store, the origin of fire, and location of the steel roll-up and man door, which served as inlets for the wind. This entire side was pressurized by a 15 mph wind which pushed and rapidly spread fire into and eventually throughout the entire length of the structure. Windows on the A side, which were broken to provide ventilation, unknowingly served as vent points for the wind-driven fire. (Courtesy of Paul Ramirez.)

The Phoenix incident analysis

The firefighters most likely based their decision to initiate a fast and aggressive interior attack on the observation of light smoke conditions showing on the interior of the store, suggesting that the fire had not been burning for a long period of time, and on the perceived degree of danger associated with an unoccupied commercial structure. As in the case of other large enclosed structure fire operations in cities across the country, without current knowledge of the extreme risk associated with this type of structure fire and the ineffectiveness and danger of an offensive strategy during wind-driven fire conditions, the firefighters became unknowing participants in a rapidly developing firefighter disorientation sequence.

13 Firefighter Disorientation Case Reviews: Considering Enclosed Structure Tactics

The Phoenix incident with enclosed structure tactics

Given that, on arrival, heavy smoke was showing from the B side and the location of the seat of the fire was generally known, the K (known) guideline would be implemented. This guideline calls for a quick attack of the visible fire while preparations are made to conduct a short interior attack if possible.

With adequate training, staffing, and equipment and an awareness of the hazards associated with wind-driven fires involving large unprotected enclosed structures, every department is capable of reducing the risk before the fire. As in the case of the Carthage incident (and every incident), some of the most important size-up factors are the wind speed and direction in relation to the location of the fire and possible vent points within the structure. This is critical information upon which tactical decisions will be based.

In this regard, firefighters should not advance into a structure through a door that would serve as a vent point without first controlling the inlet, because the fire could become pressurized on arrival or at some point during the course of the incident. Given the size-up factors typically observed during these scenarios and their association with opened structure fires, this type of incident may (incorrectly) seem to indicate a traditional attack from the unburned side. The consequences of this action, however, may involve sudden disorientation, entrapment, and incapacitation of firefighters in the dangerous flow path.

In this case, risk reduction must begin at the dispatch office, which should be given the responsibility for monitoring and transmitting current and future wind speed and directions to firefighters on dispatch of every reported structure fire. Working from previously obtained lists of critical structure information collected from line officers and firefighters, an additional notification from the dispatch office of the presence of lightweight trusses in the ceiling space of a structure that is large, unprotected, and enclosed would also be beneficial and factored into the strategy. Additionally, all responding companies would then follow steps outlined in a wind-driven fire action plan.

Common wind-driven fire scenarios involve vented fires on an exterior side of a structure that are being pressurized by a 10 mph or greater wind speed. During the Phoenix incident, the fire was pressurized by a south wind at 15 mph entering the single and roll-up doors on the B side. As in the case of the Carthage fire, the Phoenix incident was a complex and extremely dangerous incident because, in addition to all of the dangers associated with large enclosed structure fires, it was also a structure fire that was driven by the wind.

Although light smoke was showing on the interior at the Phoenix incident, heavy concentrated smoke was also initially showing from the location of the trash container on the B side of the structure. Therefore, an assumption would have been made that the seat of the fire, in this case, extension of the fire, was on the B side. With an advanced notification of a 15 mph wind speed and southerly direction obtained from the dispatch office, the first-arriving officer would have determined that the south or B side was being pressurized. If fire extension into the structure occurred, the officer would also know that the interior fire would continue to be pressurized as long as one or both doors remained open.

Assuming that the responding firefighters are fully trained in the responsibilities of safely managing wind-driven structure fires and implementing large enclosed structure SOGs, and that they are provided with appropriate equipment to safely attack enclosed structure fires and utilize handlines, portable monitors, and any hose evolution that quickly maximizes the flow, there are three tactical options firefighters could select from when attacking large enclosed wind-driven structure fires, as described in the following sections.

Option 1

In keeping with enclosed structure guidelines, and by implementing the double attack-supply evolution, the first two arriving engine companies would extinguish the dumpster fire and begin an attack on the fire that has extended into the structure. This would, however, be initiated using 200-gpm handlines to dump the booster tanks of water from each attack engine while a supply was in progress. Immediately thereafter, portable monitors would be placed on the interior of the structure with the intention of maximizing the number of portable monitors in use to

13 Firefighter Disorientation Case Reviews: Considering Enclosed Structure Tactics

four, if necessary to achieve a quick knockdown of the fire from the B side. However, this must be accomplished while keeping the wind out of the building, because if wind was allowed in the building, it could push the fire to follow the path of least resistance through unobstructed interior flow paths leading to a vent point.

In following steps of a wind-driven fire action plan, the front doors on the A side must be closed and windows left intact to prevent the creation of vent points for the driven fire to exhaust from. Additionally, truck crews should ensure that the doors on the B side remain closed, with the exception of the door to be used for charged handlines and or portable monitors. If possible, truck companies should cover the upper portion of the door opening to prevent as much air as possible from entering and pressurizing the interior.

Flows from master steams such as portable monitors rated at 500 gpm are significant. Company and sector officers should therefore monitor the smoke and conversion steam conditions to determine when to shut down the monitors and, when safe, to make a visual check of the interior and complete final extinguishment.

Option 2

While keeping with enclosed structure guidelines, and by implementing the double attack-supply evolution, the second option involves knocking down the trash fire, and after determining that extension of fire has occurred, immediately closing all doors on the B side. It would be safest to position a radio-equipped firefighter by these doors to ensure that they remained closed during the entire incident and to immediately inform command should a breach of this side occur. Additionally, another view of the roof by another radio-equipped firefighter, either from an elevated platform or aerial or ground ladder adjacent to the structure, would make sure that the roof remained intact to prevent an influx of air from above. Should a partial collapse or opening in the roof take place, command would immediately be notified.

Thereafter, the front doors and windows would be closed and need to remain closed and intact to prevent the creation of vent points. The action plan would then involve extensive pulling of ceiling tiles at the adjoining Ace Hardware Store to check for fire extension above, with the intention of breaching the common wall that separates the rear storeroom from the hardware store.

Multiple positive pressure ventilators (PPVs) would be used at the hardware store's main entrance, and handlines advanced and standing by while a breach is made. After the opening is made through the wall, portable interior monitors with semifog or full fog settings would then be advanced and positioned in the storeroom to attack the fire where seen, possibly toward the B side of the rear storeroom.

Again, sector officers would determine the effectiveness of the attack by observing the formation of conversion steam or the continuation of dark smoke, which would indicate that continued water flow is required. When appropriate, portable monitors would be shut down, and when safe to enter, firefighters would be allowed to complete final extinguishment and ventilation.

Option 3

Use of option 3, which is an advanced enclosed structure tactic, would seem to contradict avoiding the quick and aggressive interior attack unsuccessfully used by firefighters in the past. The major difference between the action that took place during past disorientation fires and those suggested here involves a greater knowledge of and understanding of the problem and avoidance of the hazards during execution of the tactic.

The use of this option therefore requires the specific avoidance of life-threatening hazards that lead to firefighter disorientation and use of the following safety control measures, which are presumed to have been incorporated into departmental training programs and SOGs:

    a. Recognize and avoid the danger associated with wind-driven fire.

    b. Utilize evolutions that provide a quick backup and quick redundancy and that quickly maximize the flow.

c. To maximize the flow, use high-flow (500-gpm) portable interior monitors with low-angle attack capability early in the incident.

(Two portable monitors assigned to each engine would allow the flow to be quickly maximized by each.)

d. Maintain company communication, integrity, and accountability.
e. Use TICs within limitations to provide orientation and reorientation for those firefighters who become separated from the handline.
f. Avoid development of loops or entangled hose in the structure.
g. Use a short interior attack whenever possible.
h. Exit the structure prior to low-air alarm activation.
i. Understand the longer time requirement for crew evacuations during deteriorating or PZVCs.
j. Obtain dispatch notification when it has been determined that all civilians have exited the fire structure.
k. Obtain dispatch notification of wind speed, direction, and changes.
l. Obtain dispatch notification of involvement of unprotected enclosed structures, basements, trusses, and structures too dangerous for an offensive or short interior attack.

While in keeping with enclosed structure guidelines, and by implementing the double attack-supply evolution, the third option would involve an immediate and safe interior portable monitor attack on the fire from the unburned side.

During this action plan, after attack and extinguishment of the dumpster fire and determining that extension of fire has occurred, all doors on the B side would be immediately closed to prevent the pressurization of the interior. It would be safest to position a firefighter at these doors to ensure that the doors remained closed during the incident. At the front, or A side, of the structure, a crew with a TIC and safety directional arrows attached to the handline would advance a 1¾-in. handline to serve as a lifeline and provide protective cover while an inspection was made. Observing a working fire in the rear storeroom through the opening situated a safe distance from the seat of the fire, other firefighters would advance and immediately charge each of four portable interior monitors (each flowing at 500 gpm × 4 = 2,000 gpm) with nozzles placed on straight or semifog setting, as needed for reach and effectiveness. If needed, additional portable monitors would be advanced by setting up additional attack pumpers at the scene, using the double attack-supply evolution or dual pumping, for a potential combined fire flow from all four attack pumpers of 4,000 gpm via portable interior monitors .

The interior streams would be aimed at either the main body of fire and or the ceiling in the rear storeroom, to ensure that extension of fire into the main floor area would not occur. The interior firefighters would be directed to leave the structure while allowing the master streams to flow unattended. Next would be to check for the production of black smoke or conversion steam to determine when it would be safe to shut the monitors down and, if structurally sound, allow firefighters to reenter to check interior conditions and complete final extinguishment, if necessary.

If this or any of the preceding options were determined to be too difficult to initiate or were ineffective, or if the safety of firefighters was in question, command would have to take quick and decisive action. For lifeline purposes, command must initially allow all interior hoselines to remain in the structure, conduct an emergency early evacuation of all firefighters, call for a PAR, and establish safe collapse zones while transitioning to a defensive attack.

The Charleston Incident

On June 18, 2007, Charleston, South Carolina, firefighters experienced the worst case of firefighter disorientation caused by PZVCs ever recorded in the United States, which resulted in the loss of nine firefighters. As in the case of other large unprotected enclosed structure fires, the Charleston Sofa Super Store fire was complex and extremely dangerous in many ways (fig. 13–10).

13 Firefighter Disorientation Case Reviews: Considering Enclosed Structure Tactics

Fig. 13–10. The 2007 Charleston Sofa Super Store fire. (Courtesy of Stewart English.)

With the benefit and heavy reliance of clear details provided by outstanding investigatory reports completed by agencies including the Bureau of Alcohol, Tobacco, Firearms and Explosives (ATF); NIOSH; the National Institute of Standards and Technology (NIST); the Phase I and Phase II Reports of the City of Charleston Post Incident Assessment and Review Team; and key information provided by members and officials of the Charleston Fire Department, this chapter presents a review and brief analysis of the strategic decisions made during the operation and proposed enclosed structure tactics for similar operations everywhere.

Fire scene photos were provided by NIOSH, the Pictometry International Corporation, and accommodating citizens help orient the reader to such information as structural dimensions, configuration, initial location of the fire, direction of the wind, and apparatus positioning on the scene prior to discussion (figs. 13–11, 13–12, 13–13, and 13–14).

Fig. 13–11. Aerial photo taken in March 2007 by Pictometry International Corporation. The annotated photo indicates the location of Engine 10 and Engine 11, supply lines, and hose lines pulled at different times during the incident. Note accumulation of trash at loading dock on the day photo was taken, 3 months prior to the incident. Note the absence of ventilation ductwork or other roof penetrations over the showroom, thus no path for smoke and hot gases to escape. Note the location of Pebble Road visible along the C side of the warehouse. The wind was variable from the south-southwest (upper right to lower left of the photo) up to 11 mph. (Courtesy of NIOSH F2007-18; see NIOSH 2009a. Reprinted with the permission of Pictometry International Corp. Copyright 2007, Pictometry International Corp. All rights reserved.)

13 Firefighter Disorientation Case Reviews: Considering Enclosed Structure Tactics

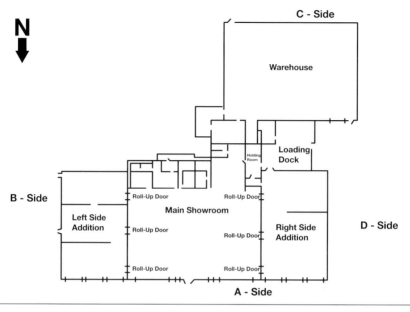

Fig. 13–12. Floor plan of furniture store and warehouse. (NIOSH F2007–18; see NIOSH 2009a.)

Fig. 13–13. The Sofa Super Store structure measured 150 ft. × 240 ft. × 25 ft. Fixed windows were situated only along the A side. The remaining exterior sides were constructed of concrete block walls or sheet metal siding. Sealed and obstructed means of egress, which prevented an emergency evacuation, were located along the perimeter of the structure. (Courtesy of Stewart English.)

Preventing Firefighter Disorientation

Fig. 13–14. The earliest photo of the front of the Sofa Super Store, taken at approximately 19:16. Assistant Chief, Engine 10, and Engine 11 were on the scene at 19:11 hours. Ladder 5 was on scene at 19:12 hours. (Courtesy of Lindsay Ackermann.)

During the early stages of the Charleston fire, and with a variable south-southwest wind speed of 11 mph, heavy smoke and fire were visible from the rear (D side) of the structure. Fire involving a pile of trash extended into the loading dock by means of an adjacent open window. An initial attack with a 1½-in. handline by the first-arriving engine company was then ordered, and the handline advanced into a rear enclosed loading dock area. Due to deteriorating conditions in the 2,200 square foot room, in which at one point, fire engulfed the interior crew, the crew evacuated, and the attack was subsequently made from the exterior in this area.

During this effort, entry was made from the A side. An investigation of the interior found it to be clear, with the exception of light smoke seen at the ceiling at the right rear portion of the store. A set of double doors that separated the uninvolved main showrooms from the involved rear loading dock was opened, and a working fire was visible, but due to the fire's powerful suction for additional air, the doors could not be reclosed.

Immediately, a deep interior attack was ordered and handlines advanced from an attack engine through the front doors on the A

13 Firefighter Disorientation Case Reviews: Considering Enclosed Structure Tactics

(unburned) side of the structure. Applying the National Fire Academy flow formula: (length × width divided by 3) × percent of involvement, a minimum of 800 gallons per minute (gpm) of water would have been required at the loading dock (NIOSH 2009a).

The interior attack was initiated with a 1½-in. handline at 90-gpm flow. A 2½-in. handline and a 1-in. booster line were also advanced. Water was delivered initially from the engine's booster tank, in an attempt to prevent fire extension into the main showroom areas of the structure. As it turns out, only intermittent flow was achieved from the 1½-in. handline, and no flow ever delivered through the 2½-in. handline.

Note that it required 15 minutes for the attack pumper on the A side to obtain a water supply delivered initially from a long lay of single 2½-in. supply line with limited overall supply capability. On the interior, additional sections of hose were added for reach, resulting in excess hose causing the formation of dangerous loops and entangled handlines.

As conditions deteriorated, heavy smoke filled the interior from ceiling to floor, resulting in PZVCs. At this critical time, a rescue operation was underway from the exterior on the C side of the structure, as other interior firefighters simultaneously began to send emergency transmissions for help in exiting the structure. The calls were not immediately heard. Nine firefighters working on the interior of the structure at that time, without the benefit of a TIC or an untangled handline, became disoriented and depleted their air supplies as they attempted to reach a means of egress (NIOSH 2009a). The dangerous distance firefighters advanced into the structure immediately after arrival and other interior conditions were specifically noted in the Charleston Phase II Report as follows:

> The area where the crews were operating was approximately 200 feet inside the showroom from the front entrance. The entire showroom was filled with smoke and the path back to the front entrance was a series of narrow aisles among the furniture displays. The only guide available to the firefighters was to follow the hose lines back toward the entrance; they had to feel their way along the hose lines which included loops and turns and

had become entangled with the furniture. If the firefighters waited until the low pressure alarms on their SCBA [self-contained breathing apparatus] activated, they would have had only 3 to 4 minutes to find an exit from the depths of the Sofa Super Store before their air supplies were exhausted. A firefighter who was disoriented or had lost contact with the hose line would have been unlikely to find a way out of the building within the limited available time. (Routley et al. 2008)

During this rescue operation, conditions quickly deteriorated as rapid fire spread caused the interior of the store to become heavily involved precluding rescue of nine firefighters, although successful rescues of several other disoriented firefighters were immediately made at great risk to the rescuers during the incident. According to the medical examiner's report, the firefighters sustained injuries caused by smoke inhalation, which caused their fatalities. As noted by the Phase II Report, these were injuries typically associated with disorientation (Routley et al. 2008).

The rapid fire development likely evolved shortly after fresh air entered the store through the front windows following the breaking of glass to ventilate for greater visibility, and an unfortunate and untimely partial collapse of the sheet metal cover over the rear loading dock area. As companies transitioned to a defensive attack, the fire extensively exposed lightweight steel trusses in the concealed ceiling space, which caused the roof to become weak. The roof then collapsed over major sections of the showrooms. (Routley et al. 2008)

The Charleston incident analysis

Prior to a discussion of enclosed structure tactics, firefighters first need to understand their tactical capabilities within acceptable levels of risk. The following important passage from the Charleston Phase II Report clearly addresses what would have been acceptable and unacceptable levels of risk during the Charleston incident.

13 Firefighter Disorientation Case Reviews: Considering Enclosed Structure Tactics

The IAFC [International Association of Fire Chiefs] Acceptable Risk Guidelines would support the initiation of an offensive attack during the initial stage of the Sofa Super Store incident, if the Incident Commander believed the fire could be contained to the loading dock without exposing firefighters to excessive risk. (This situation would be classified as *medium risk* and *marginal probability of success*.) The fire that was burning in the loading dock presented a significant tactical challenge. The fire involved a relatively large space (approximately 2,200 square feet) that was filled with highly combustible contents. A successful offensive attack would have to deliver sufficient fire flow (water) to overcome the volume of fire within this space. The situation was greatly compounded by the circumstances. Access to the fire area was difficult and the building configuration created immediate exposures on three sides. In addition to delivering a powerful attack to suppress the fire within the loading dock, the Incident Commander would have to ensure the fire did not extend into any of the exposures. The Incident Commander was responsible for determining whether the available firefighting resources had the ability to control and/or contain the fire and whether this action could be accomplished safely. The risk assessment should have changed as additional information was obtained and fire conditions were reevaluated. If the Incident Commander lacked the capability to conduct a safe and effective offensive fire attack in the time that was available, the strategy should have changed to defensive. As soon as the fire extended into the void spaces above the showrooms, the situation exceeded the capability of the Charleston Fire Department to control the fire with an offensive strategy. Multiple large hose lines would have been required to stop the spread of hot fire gases within the void spaces. The hose lines would have to be operated by crews operating inside the showrooms and opening ceilings to attack the fire. This attack would have to be coordinated with vertical ventilation, opening holes in the roof to release the trapped smoke and fire gases. The Charleston Fire Department did not have the resources, training, or leadership that would have been required to conduct an operation of this size and complexity in the limited time that was available. A risk management analysis at that point would have determined that attempting to conduct an interior offensive fire attack under these circumstances place firefighters in conditions of unacceptable risk. (The risk analysis would classify this situation as *high risk* and *low probability of success*.) The revised risk analysis would dictate a switch to defensive strategy and the withdrawal of all firefighters from interior positions. (Routley et al. 2008)

The Charleston incident with enclosed structure tactics

Since a heavy column of smoke was showing from the rear of the structure along the D side when the first companies arrived on the scene, they would have known where the fire was located. Therefore the K guideline would have been implemented. During the execution of this guideline, visible fire is immediately attacked while preparations are made to implement a short interior attack. However, the wind speed and direction must always be taken into consideration and factored into the operation (fig. 13–15).

With an awareness of the hazards associated with large, unprotected, enclosed structure fires under wind-driven fire conditions, departments are capable of reducing the risk before the fire in many ways. Should a similar fire occur in any community today, a different approach could be used by departments staffed, trained, and equipped to do so. However, those not properly prepared should not attempt these tactics, as ineffective and unsafe attempts may result in injury to firefighters. Due to resource limitations, these departments would instead concentrate on protecting exposures while engaging in safer defensive operations.

13 Firefighter Disorientation Case Reviews: Considering Enclosed Structure Tactics

Fig. 13–15. Note the angle of the smoke plume which corresponds roughly to the 11 mph south-southwest variable wind speed reported during the incident. (Courtesy of Dan Folk and NIOSH F2007-18; see NIOSH 2009a.)

Light, moderate, or heavy smoke showing from an enclosed structure tells firefighters that they are dealing with an extremely dangerous situation in which firefighters may be exposed to multiple life-threatening hazards.

During an incident like the one in Charleston, the life safety hazard would have been initially addressed. During the Sofa Super Store fire, all employees with the exception of one had exited the structure before the first company arrived. Given that heavy smoke was seen at the rear D side, the seat of the fire would have presumably been located in that portion of the structure. The fire involved trash on the exterior with the possibility of extension, which suggested that the fire had not been burning for a long time at the point when firefighters were arriving. A 360° size-up would have been conducted, but due to the large size of the structure, it would have been delegated and conducted by another trained and subsequently arriving officer, firefighter, or battalion chief's aide. The information gathered from the walk-around would have then been provided to command.

Option 1

In keeping with the overall safety measures advocated by enclosed structure guidelines, option 1 would call for a quick attack of the Sofa Super Store fire, which was rising above the side of the structure next to the loading dock doors and extending into the loading dock area, by the first two arriving engine companies using the double attack-supply evolution. The objective here is to discharge on the fire the entire volume of water from both booster tanks using 1¾-in. handlines flowing 200 gpm while the third engine supplies (fig. 13–16). Other responding companies would be directed to prepare for a short interior portable monitor attack.

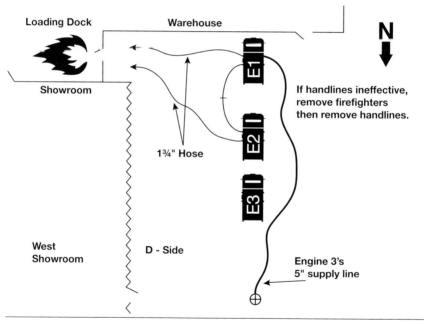

Fig. 13–16. Option 1: Using the double attack-supply evolution, both engines lay 1¾-in. handlines to quickly discharge all the water in their booster tanks, while preparations are made for a short interior portable monitor attack. Total initial flow = 400 gpm.

Option 2

Option 2 would call for quickly attacking the fire using one portable monitor from each of the arriving attack engines as the third company supplies (fig. 13–17). Other arriving companies would be ordered to prepare for a short interior attack.

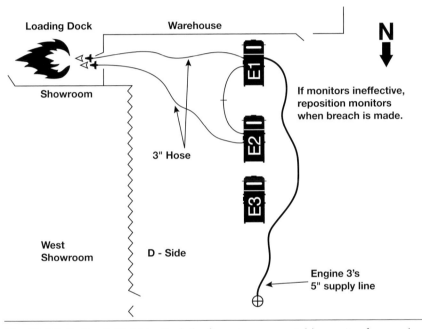

Fig. 13–17. Option 2: Quickly attack the fire using one portable monitor from each of the first two arriving engine companies. Total flow = 1,000 gpm.

This attack with monitors is also executed with the objective of quickly dumping all of the booster water from both attack engines onto the fire. It is understood, however, that when flowed from the booster tank, a 500-gpm monitor will discharge the water from a 500-gpm booster tank in about 1 minute, 1.5 minutes from a 750-gallon tank, and 2 minutes from a 1,000-gpm tank. It should be widely understood by command, engine company officers, and all firefighters that this must be factored into the tactic to ensure a safe attack by ensuring prompt firefighter egress.

It is conceivable in some cases involving reduced volumes of fire that these types of initial attacks, using either two 1¾-in. handlines (one 1¾-in. handline from each of the attack engines) or two portable monitors (one interior portable monitor from each of the attack engines), may be effective in knocking down the main body of fire, which would allow companies to complete final extinguishment as a supply is established. In this case the short interior attack would therefore not be necessary.

Note, though, that these attacks would have required a fire curtain to be placed over the opened window as soon as possible. This would be done to prevent air from entering the rear enclosed loading dock area, as attack lines or portable interior monitors were used through the single sliding door that led to the heavily involved room. In cases where the initial attacks described in options 1 and 2 were not effective, companies would exit and, after making sure that all the firefighters were out, remove the handlines and monitors from the structure. After a breach was completed, the monitors would then be repositioned at an advantageous location on the interior to cut off the spread of fire. To prevent air from feeding and driving the fire toward the structure's interior, firefighters would have to remember to close the sliding door to the rear loading dock room and keep the curtain on the window. In addition, it would be best if a radio-equipped firefighter were placed at this position to ensure that the openings remain closed during the course of the operation.

Option 3

Using the double attack-supply evolution and in an effort to cut off the spread of fire, four 500-gpm rated portable monitors, two from each attack engine, each fed by a single 3-in. line, would be used initially to provide a combined flow of 2,000 gpm (fig. 13–18). If monitors initially used in the loading dock room were ineffective, they would need to be repositioned after the breach was made. If more water was required, another double attack-supply evolution or a dual pumping evolution by various means would be carried out (fig. 13–19).

13 Firefighter Disorientation Case Reviews: Considering Enclosed Structure Tactics

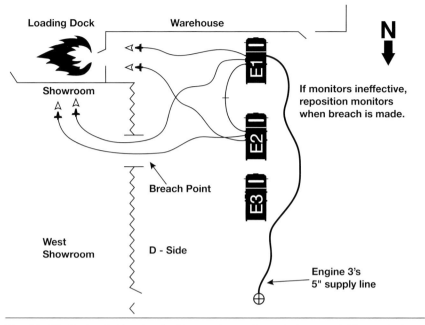

Fig. 13–18. Option 3: Use the double attack-supply evolution to quickly attack the fire by using two portable monitors from each of the two initially arriving engines. Total flow = 2,000 gpm.

Option 4

By setting up another double attack-supply evolution or dual pumping evolution such as the attack-supply evolution with dual pumping at the scene or the single-engine evolution with dual pumping at the scene, a total combined flow of 4,000 gpm provided specifically from two 500-gpm interior portable monitors, from each of four attack pumpers (8 × 500 gpm = 4,000 gpm), can be achieved for operations involving the protection of the structure's west and main showrooms. Additionally, should Engines 5 and 6 be needed to provide more flow using interior portable monitors as established by the earlier arriving engines, the total combined flow for the operation would be increased to 6,000 gpm (fig. 13–19). Notice that in this case, in an effort to quickly maximize the flow, a 2-to-1 ratio of engines to supply lines is established, with each 5-in. hose supplying two engines and, if necessary, three 5-in. large diameter hoses supplying six engines.

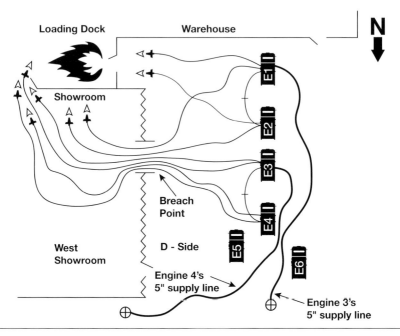

Fig. 13–19. Option 4: This option calls for using two dual pumping operations with four attack engines, each pumping for two portable monitors, eight portable interior monitors in all. This flow could be delivered from several key interior positions to help cut off the spread of fire by executing various available dual pumping evolutions. Total flow = 4,000 gpm. If needed, Engines 5 and 6 could be placed into operation to increase the total flow to 6,000 gpm. (Figure not to scale.)

In reference to tactics depicted in figure 13–19, if the first two monitors placed into operation by Engines 1 and 2 at the loading dock are ineffective, reposition them in the showroom when the breach is made. In addition, with the evolution choices available, note the flexibility gained as the double attack-supply evolution is immediately executed with Engines 1, 2, and 3 while the attack-supply evolution is executed with dual pumping at the scene with Engines 3 and 4. It is interesting to also observe that in the two previously executed evolutions, Engine 3 served dual roles: first, as the supply engine during execution of the double attack-supply evolution, and second as an attack engine during subsequent execution of the attack-supply evolution with dual pumping at the scene. If more fire flow is needed, the single-engine evolution may be executed with dual pumping at the scene with Engines 5 and 6 or, the attack-supply evolution with dual pumping at the scene may be executed

13 Firefighter Disorientation Case Reviews: Considering Enclosed Structure Tactics

with Engines 5 and 6. In addition to the execution of these hose evolutions, other tasks crucial to the operation include the following:

- A radio-equipped firefighter should monitor the condition of the roof and immediately inform command if signs of weakness are visible or a partial or complete collapse of the roof occurs.

- The front doors and windows on the A side should remain closed, intact, and guarded by another radio-equipped firefighter. This would prevent creating vent points for fire if wind makes entry at any time during the fire from the C (south) or C-D (southwest) side.

- A short interior portable monitor attack, if each company used TICs, would be quickly undertaken, taking advantage of the mobility and low-angle capability of 500-gpm rated monitors connected to single 3-in. feeder lines from positions required within the structure.

- Portable monitors that achieve a greater flow than discussed here can be used at individual department discretion; however, safe discharge at lower angles may not be achievable.

- Engine operators must also ensure that the full capacity of their engine is not exceeded by the total number of monitors ultimately supplied by that engine. For example, the capacity of the engine would be exceeded if the operator tried to supply two 800-gpm tips for a flow of 1,600 gpm from an engine rated at only 1,000 gpm or 1,250 gpm.

- During preparations for a short interior attack at an incident similar to the Charleston fire, and with the help of employees who were present on the scene, firefighters would cut through the structure's sheet metal siding at a point that would avoid interior contents from blocking firefighters from entering the breach point.

- The large breach point, created on the exterior D side wall and wide enough to allow the simultaneous entry of at least two firefighters, side by side, would be selected to provide a short distance to travel into the area of the fire and also to achieve a safe, rapid exit without delay due to congestion if needed.

Firefighters must also consider controlling the wind entering this opening.

- Once the large opening in the wall is made, firefighters must pull the ceiling to scan the ceiling space in search of smoke, fire, or unprotected steel, which may elongate during fire exposure and cause a collapse of the roof, and for lightweight steel or wood trusses, which may also suddenly collapse when exposed by high heat or fire.
- The location of the breach point should also allow the placement of the monitors to effectively reach, attack, and cut off the fire involving the rear enclosed loading dock room.

Specifically, and with the advantage of the passage of time, hindsight, and analysis, monitors could be placed just outside the double doors that lead into the showroom and or just inside the enclosed loading dock room. Nozzles would be set to either full fog, semifog, or straight stream patterns as deemed most advantageous by the officer and aimed directly at the fire or deflected off the ceiling and allowed to drop onto the fire.

When it is not possible to approach the fire, keep in mind that monitors currently available can flow 500 gpm a distance of 200 ft. Therefore, officers should always take advantage of the capability and safety associated with a monitor's reach.

Because portable monitors are not typically charged prior to positioning on the ground, they will be charged immediately after placement, which will therefore charge 2½-in. or 3-in. feeder lines to serve as life lines out of the structure. Charged 1¾-in. handlines for protective cover of interior firefighters or for additional attack purposes can be advanced as needed or as determined by the engine officers or command.

Again, with the benefit of hindsight, after the first two monitors are in place, with the intention of ultimately advancing and positioning more backup master stream appliances in the structure, additional ceiling tiles must be pulled by truck crews at the common wall separating the right rear showroom from the rear involved enclosed loading

13 Firefighter Disorientation Case Reviews: Considering Enclosed Structure Tactics

dock area. If fire is visible along the top of the wall or through the wall, a monitor should be charged and visible fire attacked.

One nozzle should be aimed at the double doors before they are opened, and when they are opened, the master stream should be discharged into the loading dock room. The presence of lateral extension of fire through the ceiling (void) space above the holding room next to the loading dock area must also be determined by aggressively pulling additional suspended ceiling tile and deploying additional portable monitors to attack the fire seen above.

As in the case of the Memphis incident, the doors immediately separating rooms involved in the fire should not be opened until charged attack line(s) or portable interior monitor(s) capable of controlling the volume of fire are standing by and ready for immediate use.

Recall that some types of portable monitors are capable of safely lowering the stream downward a few degrees above horizontal to allow a heavy attack to be made from interior positions through doorways when located either up close or at a distance from the target. This ability is advantageous in this setting, as it allows for an interior monitor attack (500 gpm) onto larger volumes of fire while maintaining protection from heavy fire spread at floor level. Conversely, because of the lower firefighter-to-gpm ratio, the volume from multiple handlines, which would require greater staffing of interior firefighters, may not provide the required knockdown during interior attacks in large enclosed structures. However, multiple, well-positioned, interior master streams in the form of portable monitors may. This greater fire flow may also be developed by fewer firefighters in a shorter amount of time.

Another complexity associated with the Charleston incident involved the simultaneous required rescue of a trapped employee, which ultimately was accomplished from the exterior and with the assistance of the trapped employee and in close coordination with the dispatch office. During this scenario, another available company or companies working under the direction of a sector officer would be given this rescue assignment. Additionally, extension of fire into the large storage warehouse situated to the far south (C-D) corner would have been sectored and, if safe, attacked by use of multiple portable interior monitors from doors or other breach points ahead of the fire spread along the perimeter of the warehouse by additional attack pumpers set up and supplied from the two hydrants available and located immediately behind the warehouse on Pebble Road.

Fires in large unprotected enclosed structures are complex and have the potential to present any of a number of challenges, some of which may be insurmountable. Large enclosed structures are also extremely dangerous, having taken the lives of multiple firefighters at single fire events, with and without the introduction of the wind factor (figs. 13–20, 13–21, 13–22, and 13–23). Therefore, if conditions indicate that safety is being compromised and are recognized by any firefighter on the scene, in this case, whether in the main showrooms or rear warehouse, an early emergency evacuation from the structure should be made, attack lines advanced into the structure left in place until a PAR is completed, and a transition to a defensive attack with collapse zones initiated.

Fig. 13–20. Showroom windows are broken for visibility. (Routley et al. 2008; photo courtesy of Stewart English.)

13 Firefighter Disorientation Case Reviews: Considering Enclosed Structure Tactics

Fig. 13–21. "There is ample evidence that breaking the windows provided air to the fire and accelerated the ignition of the showroom contents." (Routley et al. 2008; photo courtesy of Stewart English.)

Fig. 13–22. "The windows were broken at approximately 19:35." (Routley et al. 2008; photo courtesy of Stewart English.)

Fig. 13-23. "The main showroom became fully involved within three to four minutes." (Routley et al. 2008; photo courtesy of Stewart English.)

Dual Pumping at a Distant Hydrant with 5-in. Hose

After ensuring that life safety has been addressed, the next tactical priority during large enclosed structure fires is to quickly maximize the flow to control a potentially large fire. A specific tactical problem with fireground hydraulics developed during the early stages of the Charleston Sofa Super Store fire, and because it may be encountered in any department, it deserves a brief discussion here. This discussion is by no means a replacement for a comprehensive engine operations course, including proper use of departmental apparatus and flowcharts, but rather focuses on one specific and common fireground issue.

At the Charleston incident, a nearby hydrant had been removed prior to the fire, so a supply engine had to lay and connect to a more distant hydrant to provide a water supply for the attack engine positioned on the A side of the structure. The distance to the hydrant was 1,850 ft.

13 Firefighter Disorientation Case Reviews: Considering Enclosed Structure Tactics

However, because a single 2½-in. supply hose was used and the hose lay was long, the excessive resulting friction loss prevented an effective flow from developing. This long hose lay requirement was clearly documented by the Charleston report. This is a situation experienced by firefighters throughout the fire service utilizing various diameters of supply hose and thus firefighters need to understand the capabilities and limitations of the hose, appliances, and apparatus used on the scene.

One solution offered for suitably equipped engines, assuming that 10 sections of 5-in. hose are carried in all hosebeds, is to lay a 5-in. supply line to the distant hydrant, get filled in due to laying short, and pump the 5-in. line to an attack engine at the scene. If the fire indicated that another attack engine would be needed, the evolution would be repeated. In this case, another supply engine would lay toward the engine on the hydrant, get filled in due to laying short, and connect to the engine on the hydrant in a dual connection using a 5-in. inlet-to-inlet soft suction hose connection. The second supply engine operator would then pump the 5-in. supply line laid to a second attack engine at the scene (fig. 13–24).

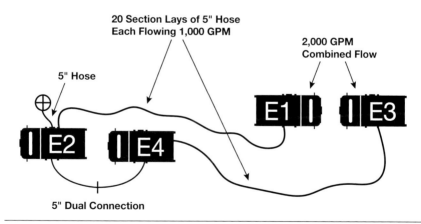

Fig. 13–24. This evolution illustrates dual pumping from a distant hydrant with 5-in. hose. In this evolution, a total of 2,000 gpm, (1,000 gpm from each engine) is delivered from a distant hydrant through two 20-section lays of 5-in. hose. Each lay = 2,000 ft. Two additional engines would each lay 10 sections of 5-in. hose to fill in the short hose lays. Engine numbers correspond with the order of arrival of pumping engines.

Because of the greater distance to the hydrant, maximum flows may not be developed but dual pumping from a distant hydrant will at a minimum ensure that adequate flow from the supply engines is available for the attack engines to use. As always, for safety, operators should gradually charge 5-in. supply lines to avoid damage that may be caused by water hammer. Additionally, the supply lines would both be pumped using the engine pressure obtained using the formula:

Engine pressure = friction loss + 10 psi (for fluctuation)

Note that engine operators should add or subtract ½ psi for each foot of elevation increase or decrease, respectively, to ensure that pressures and flow delivered are reasonably accurate.

If, for example, on a level grade, 1,000 gpm is required for one of the attack engines, with a corresponding friction loss of 8 psi per 100-ft. section, and 20 sections were laid, the engine pressure set by the supply engine operator would be 170 psi. This pressure is obtained by calculating the following: Friction loss = 8 psi per section × 20 sections of 5-in. hose = 160 psi + 10 psi for fluctuation = 170 psi in volume.

For engines with two-stage pumps, volume settings on transfer valves are used when an engine attempts to flow more than 70% of its rated capacity. In this case, a 1,000-gpm supply engine trying to deliver 1,000-gpm flow would be attempting to deliver 100% of its rated capacity and therefore would be within the volume setting.

With respect to the original distant hydrant evolution, 170 psi is a safe engine pressure that will result in a flow of 1,000 gpm through 20 sections of 5-in. hose. The operators at the attack pumpers at the scene will then be capable of setting and maintaining required engine pressures for the water flow in use while also avoiding cavitation, which is caused by attempting to pump more flow than available and which may damage the pump.

As long as the engine operators at the scene maintain at least 10 psi incoming residual pressure and the engine operators at the hydrant work at or below the working pressure limit of 5-in. hose, this evolution will perform safely and effectively. The working pressure limit of 5-in. hose is approximately 180 psi (this varies a bit by manufacturer).

Due to the effect of friction loss or the pressure loss realized as water flows through hose, increasing the flow (e.g., 1,000 gpm) generally reduces the distance water can be delivered. Reducing the flow (e.g., 800 or 500 gpm) increases the distance water may be delivered. For a given flow, 2½-in. or 3-in. supply hose generates greater friction loss, and 5-in. hose develops less friction loss even when attempting to deliver major flows.

Effectiveness of Vertical Ventilation

During the Charleston fire and other similar operations at large enclosed structure fires such as the Phoenix incident, the use of vertical ventilation may not have assisted significantly in improving interior conditions. Vertical ventilation works well when wind is not a significant factor; it may not be as effective during wind-driven fire conditions of 10 mph or greater in speed.

According to information in a 2010 briefing on the NIST Charleston Fire Study by Nelson Bryner, deputy division chief for the Fire Research Division of the NIST Engineering Laboratory, two vertical ventilation models were developed during the fire modeling analysis of the Charleston incident. These models used a small and a larger hole in the roof. Concerning the effectiveness of these vent holes, Bryner noted that initially there was some improvement in the interior, but no appreciable improvement in conditions subsequently (NIST 2010).

This would suggest that utilizing vertical ventilation during large enclosed structure fires with heavy fuel loading and flat roofs under wind-driven fire conditions may not provide the benefits of improved interior conditions, including visibility and tenability. To avoid needless exposure to risk, NIST recommends further studies on the effectiveness of vertical ventilation involving large enclosed structures during various wind speeds.

Additionally, company, sector, safety, and command officers must forecast what events may take place that could endanger firefighters on the scene during the ongoing size-up process. This would require timely progress reports from assigned sector and interior company officers.

During the Charleston incident, like the Carthage fire, a partial collapse of the roof over the fire in the rear enclosed loading dock area occurred. These events, which take place during ongoing operations, can rapidly make interior conditions untenable by a rush of air that is typically forced into a structure as a result of a collapse of a roof. These collapses may rapidly push the fire throughout the interior of the structure.

In addition, the prevailing wind may also force fire inward and, following the path of least resistance, out any available vent points. The speed with which wind-driven fires can travel through the interior of a structure varies, but according to firefighter witnesses at the scene of a double firefighter fatality warehouse fire in Salisbury, North Carolina, fire spread from one end of the large warehouse to the other in a matter of seconds. This followed a floor collapse that enabled fire to rise from the basement. This in turn caused the fire to rush throughout the structure, through open interior doors and exterior windows, by a 10 mph wind entering a window situated on the B side of the building (NIOSH 2009b).

Large, unprotected, enclosed structure fires are complex when wind is not a factor. However, when an unfavorable wind or induced ventilation is introduced, with the possibility of producing rapid fire spread, these should be considered the most dangerous type of structure fire that firefighters will be called upon to fight. Therefore, for these extremely dangerous operations, as much potential risk as possible must be anticipated, thoroughly planned for, and managed before the fire ever occurs and before firefighters are even permitted to make interior fire attacks into these types of environments.

In other words, the plan must be established, practiced, and well understood by every dispatcher, firefighter, apparatus operator, and every chief, company, and safety officer involved. The plan must also be unique to the resources available and updated on an ongoing basis to ensure that firefighters benefit from the very latest and safest tactical and technological resources available.

"Fire through the Roof" Tactics

At this point, it is appropriate to review a specific tactic previously discussed. During large enclosed structure fires, attacking the fire where seen applies both to fire visible on a side of the structure and fire coming through the roof. Whenever fire is showing through the roof of an enclosed structure and there is no life hazard present, quickly attack the fire seen from the exterior (fig. 13–25).

Fig. 13–25. Large enclosed structure fires are extremely dangerous and require quick developing tactics to control and extinguish. When fire is showing through the roof on arrival and life safety is not a concern, initially attack the fire where seen from safe exterior positions. (Courtesy of Dennis Walus.)

When ceiling spaces are present, the objective of this tactic is to sufficiently reduce the temperature in the ceiling space and interior in the immediate area of the fire. This will eliminate the heat required to trigger a backdraft in the presumed oxygen-depleted ceiling space or prevent an engulfing flashover from occurring.

After a reasonable application rate and time, with additional charged handlines on standby, shut down the flow from the handlines or master streams and pull the ceiling at a point of entry that is closest to the seat of the fire to see if the fire in the ceiling space has been knocked down. Be aware of the possibility that lightweight trusses or beams may have been weakened by exposure to the fire. If an assessment with a TIC determines that this is not the case, then continuously pull ceiling tiles as entry and advancement are made to expose and extinguish any remaining hot spots.

This tactic offers an alternative approach to the traditional fast and aggressive interior attack from the unburned side and the pulling of ceiling in deep interior positions, which are linked to firefighter fatalities. In the process, potential backdrafts may be avoided and potential weakening of the roof assembly, which may cause a partial or complete collapse of the roof, may also be prevented.

This safer approach may additionally avoid PZVCs, fire exposure, disorientation, and the deep and dangerous distances encountered in larger enclosed structures where firefighters would, because of total and sustained loss of vision, be literally forced to crawl through debris and entanglement to safely reach a means of egress.

There are several ways that the fire showing can be attacked from the exterior:

1. Quickly attack the fire showing through the roof with 1¾-in. handlines from the first two arriving engine companies. If fire continues to vent during this application, officers must ensure that the booster tanks from both are completely emptied as the third engine company lays the 5-in. supply line for both using the double attack-supply evolution. When practical and against a stable wall, use of ground ladders may be required to make the attack from extended ladders and not from the potentially weakened roof of the involved structure.

2. Quickly attack the fire showing through the roof with fixed vehicle-mounted monitors from the first two arriving engine companies. The engine pressure may need to be lowered for the type of fog nozzle or smooth bore tip to allow the stream to effectively lob onto the fire. The third engine company lays the 5-in. supply line for both using the double attack-supply evolution.

3. Quickly attack the fire showing through the roof with streams from an elevated platform or aerial ladder.

References

National Institute of Standards and Technology (NIST). 2010. "NIST Study on Charleston Furniture Store Fire Calls for National Safety Improvements." October 28. http://www.nist.gov/el/fire_research/charleston_102810.cfm.

National Institute for Occupational Safety and Health (NIOSH). 2000. "Six Career Fire Fighters Killed in Cold-Storage and Warehouse Building Fire—Massachusetts." http://www.cdc.gov/niosh/fire/reports/face9947.html.

———. 2002. "Supermarket Fire Claims the Life of One Career Fire Fighter and Critically Injures Another Career Fire Fighter—Arizona." http://www.cdc.gov/niosh/fire/reports/face200113.html.

———. 2004a. "Career Fire Fighter Dies Searching for Fire in a Restaurant/Lounge—Missouri." http://www.cdc.gov/niosh/fire/reports/face200410.html.

———. 2004b. "Partial Roof Collapse in Commercial Structure Fire Claims the Lives of Two Career Fire Fighters—Tennessee." http://www.cdc.gov/niosh/fire/reports/face200318.html.

———. 2009a. "Nine Career Fire Fighters Die in Rapid Fire Progression at Commercial Furniture Showroom—South Carolina." http://www.cdc.gov/niosh/fire/reports/face200718.html.

———. 2009b. "Two Career Fire Fighters Die and Captain is Burned When Trapped during Fire Suppression Operations at a Millwork Facility—North Carolina." http://www.cdc.gov/niosh/fire/reports/face200807.html.

Routley, J. G., Chiaramonte, M. D., Crawford B. A., Piringer, P. A., Roche, K. M., and Sendelbach, T. E. 2008. "City of Charleston Post Incident Assessment and Review Team: Phase II Report." http://www.dps.state.ia.us/fm/fstb/NewWebStuff2012/TrainingResources/PDFs/PhaseIIReport.pdf.

14

SUMMARY AND CONCLUSION

Unprotected enclosed structures can be located in every community of the country, and consequently volunteer and career members of every department can be exposed to life-threatening hazards. Since the hazards may be present on arrival or develop during the course of the operation, firefighters unfamiliar with the significance of a firefighter disorientation sequence may experience serious injuries or loss of life.

Unprotected enclosed structures of all types unfortunately will continue to be the firefighters' working environment far into the future. Therefore, now is the time to review current guidelines and to consider the incorporation of safer enclosed structure tactics. Additionally, it must be clearly understood that enclosed structures, unlike opened structures, are so dangerous that it is risky to be involved in enclosed structure fire operations due to their glaring fatality record. It is hoped that this trend will soon be reversed because of proactive leadership in effecting immediate or phased-in changes, as budgets allow.

The unprotected enclosed structure problem is not confined exclusively to fires taking place in the United States. In 2007 four Warwickshire, UK, Fire and Rescue Service firefighters lost their lives during an interior attack at a large commercial warehouse fire that involved a converted aircraft hangar. Two firefighters in Brussels, Belgium, lost their lives when they entered a structure and pulled ceiling to check for fire extension, triggering a backdraft that led to their fatalities.

Preventing Firefighter Disorientation

Numerous unprotected enclosed structure fires take place in the United States and abroad. However, the disorientation sequence does not always fully unfold, and firefighter fatalities are thus avoided (fig. 14–1). The threat, however, remains ever constant.

Fig. 14–1. This unprotected enclosed structure fire, which occurred in Wolverhampton, UK, measured 60 ft. × 45 ft. and involved an unoccupied indoor swimming pool facility. "On arrival no smoke or fire was showing but on internal inspection a small flame about 3" was visible coming from a vent inside the building. The Officer-in-Charge sent the crew for a hose reel and on their return the fire rolled across the ceiling, engulfing the building end to end in a matter of seconds. The building was a total loss but there were no casualties. Photo was taken from the front of an arriving appliance." (Eddie Robertson, Crew Commander, Wolverhampton Fire and Rescue; Courtesy of Wolverhampton, UK, Fire and Rescue.)

By training and routinely incorporating key steps into enclosed structure fire operations, firefighters can prevent the firefighter disorientation sequence from ever unfolding and thereby significantly increase the chance of firefighter survival. In summary, firefighter disorientation may be prevented by doing the following:

- Ensure that firefighters are able to recognize opened and enclosed structures.
- Use an accurate enclosed structure size-up.

14 — Summary and Conclusion

- Take a worst-case scenario approach when managing the incident.
- Obtain a wind report (speed, direction, and changes) from dispatchers.
- Obtain a citizen evacuation report from dispatchers.
- Obtain an enclosed structure fire involvement report from dispatchers.
- Develop enclosed structure tactics and guidelines unique to the available resources.
- Routinely use the "softening the target" method for pre-entry cooling purposes.
- Ensure that all firefighters understand their roles during an enclosed structure incident.
- Ensure that all firefighters can visually identify and understand the danger presented by various types of truss and engineered wood construction.
- Use a safe approach incorporating a cautious and calculating interior assessment that uses a thermal imaging camera for strategic and tactical decision making.
- Understand that firefighters should never risk their lives needlessly to save a structure or person already lost.
- Ensure that all firefighters can identify the life-threatening hazards that lead to firefighter disorientation and know how to take the correct actions to avoid exposure.
- Maintain company communication and integrity.
- Use modern hose evolutions that provide additional safety benefits.
- Maintain proximity to or constant contact with a handline for lifeline purposes.
- Initially allow charged handlines to remain in the structure for lifeline purposes when an evacuation order has been transmitted.

- Use safety equipment and technology such as thermal imaging cameras, safety directional arrows, advanced photoluminescence, wind control devices, high-rise nozzles, and 5-in. hose to help prevent firefighter disorientation.
- Use piercing nozzles to quickly and safely extinguish fires in enclosed spaces such as basements.
- Use appropriate forcible-entry tools and heavy equipment for safe breaching of exterior walls.
- Use coordinated ventilation in a safe and timely way.
- Utilize rapid intervention teams, who stage at the point of entry, during cautious interior assessments or short interior attacks.
- Exit involved structures prior to low air alarm activation.
- Evacuate structures early before conditions prevent safe evacuation.

Firefighter disorientation has been costly not only in terms of injuries and fatalities of firefighters and civilians, but also in terms of the enormous loss of personal property. However, by learning from the lessons of the past, with appropriate training, adequate staffing, and adoption of enclosed structure fire tactics and principles, progressive departments can permanently prevent the firefighter disorientation hazard and unwanted outcomes.

In the future, firefighters will continue to be called upon to enter structures to extinguish hazardous fires, but by avoiding as much of the predictable risk as possible before the fire, use of a safer and more methodical approach that assumes the worst-case scenario, and a slower speed for informed decision making, the manner in which attacks are made into enclosed structure fires will forever change. In the process, departments will change their tactics and will no longer allow firefighters to blindly rush, deeply and without measured aggressiveness, into structure fires associated with the repeated disorientation and fatality of firefighters (fig. 14–2).

In the final analysis, any involved structure can be considered dangerous and can take the life of a firefighter, but we now know that some are more dangerous than others. Simply stated, it is hoped through the information provided here firefighters will be safer and can significantly increase their chances of survival during the course of structural firefighting operations.

Fig. 14–2. For the prevention of firefighter disorientation, firefighters should strongly consider the use of enclosed structure tactics.

APPENDIX

Unprotected Enclosed Structures: A Global Problem

If unprotected enclosed structure fires are capable of taking the lives of multiple firefighters at single events, they can take the lives of civilians as well. More specifically, tragic fire events involving citizens or firefighters occur when an enclosed structure is not protected by an operable sprinkler or standpipe system. The following data point to the global problem, and although it is unknown if the structure fires listed actually had an enclosed structural design, their occupancy provides a powerful clue. It is also unknown whether the structures were protected by an operable sprinkler system. If they had been, however, it is reasonable to assume the outcome would not have been so unfavorable. Consequently, the enclosed structure fire problem can be regarded as tremendous and global in nature, including occupancies of various classifications.

The information in table A–1 is taken from only one of several decades of data provided by the *Our Lady of Angels Fire Memorial* website. It serves as a sober reminder of the potential threat for firefighters everywhere. Each entry includes the incident date, the location, the occupancy, and the number of civilian fatalities suffered (Morgan 2013). Some of the events were not structure fire incidents.

Table A-1. Other infamous and deadly fires from around the world

2001	3/26	Machakos, Kenya	School	68
2001	8/13	Erwadi, India	Mental home	25
2001	8/17	Quezon City, Philippines	Hotel	75
2001	9/1	Tokyo, Japan	Multiple buildings	44
2001	12/29	Lima, Peru	Shopping center	291
2002	6/26	Agra, India	Factory	42
2002	7/9	Palembang, Indonesia	Karaoke bar	42
2002	11/2	El Jadida, Morocco	Prison	50
2002	12/1	Caracas, Venezuela	Nightclub	47
2003	1/23	Tamil Nadu, India	Wedding hall	57
2003	2/20	West Warwick, Rhode Island	Nightclub	100
2004	8/1	Asunción, Paraguay	Supermarket	464
2004	9/3	Beslan, Russia	Chechen and Ingush terrorists stormed Beslan School Number One on September 1, 2004, taking around 1,200 students, staff and parents hostage. On the third day of the siege, Russian security forces stormed the school when the terrorists detonated explosives in the gymnasium, where they held most of the hostages. The explosion and resulting fire, as well as the shootout between security forces and terrorists, killed at least 334 hostages, including 186 children, and injured more than 725. At least 10 Russian security forces were killed, along with nearly all of the terrorists, bringing the overall death toll for the tragedy to at least 380.	334
2004	12/30	Buenos Aires, Argentina	Nightclub	194
2005	12/15	Jilin, China	Hospital	39
2006	2/24	Chittagong, Bangladesh	Textile factory	65
2006	12/9	Moscow, Russia	Drug rehab hospital	46
2007	3/19	Kamyshevatskaya, Russia	Nursing home	63

Year	Date	Location	Description	Deaths
2008	1/7	Icheon, South Korea	Warehouse	40
2008	4/25	Casablanca, Morocco	Mattress factory	55
2009	1/1	Bangkok, Thailand	Nightclub	66
2009	2/7	Victoria, Australia	Multiple bushfires during an extreme heatwave, fanned by winds up to 60 mph, destroyed over 5,000 structures, killing 173, injuring over 400 and leaving thousands homeless. One of the towns worst hit was Marysville, where 90% of the town was destroyed.	173
2009	6/5	Hermosillo, Mexico	Daycare center	48
2009	9/13	Taldykorgan, Kazakhstan	Drug abuse clinic	38
2009	12/4	Medan, Indonesia	Karaoke bar	20
2009	12/5	Perm, Russia	Nightclub	125
2009	1/31	Molo, Kenya	Oil spill	113
2010	6/3	Dhaka, Bangladesh	A fire that started when an electrical transformer exploded destroyed many homes.	117
2010	11/15	Shanghai	High-rise apartment building	58
2010	12/8	Santiago, Chile	Prison	81

Courtesy of Eric Morgan

Reference

Morgan, E. 2013. "Other Infamous and Deadly Fires from Around the World." *Our Lady of the Angels (OLA) School Fire, December 1, 1958.* http://www.olafire.com/otherfires.asp.

GLOSSARY

360° size-up. The collection and evaluation of fireground factors obtained from all four sides of a structure.

attack-supply evolution. A traditional hose evolution in which the first-arriving engine company attacks the fire while the second engine lays a supply line for the first.

attack-supply evolution with dual pumping at the scene. A contemporary hose evolution in which the second-arriving engine, using a straight or reverse lay, lays a supply line for the first engine at the scene. Thereafter, the second or subsequently arriving engine makes a dual connection with the first engine to obtain a supply of unused water.

backup line. A second handline used to back up the first engine company after a supply line has been established.

cautious interior assessment. A carefully executed interior size-up of conditions with the use a thermal imaging camera to determine the safety of initiating an interior attack.

dangerous. Term used to describe the degree of danger associated with an opened structure fire in which traumatic firefighter fatalities occasionally occur.

disorientation secondary to backdraft. The loss of direction due to the lack of vision caused by smoke or fire during a backdraft.

disorientation secondary to collapse. The loss of direction due to the lack of vision caused by smoke or fire during a floor or roof collapse.

disorientation secondary to conversion steam. The loss of direction due to the lack of vision caused by conversion steam produced during fire extinguishment.

disorientation secondary to flashover. The loss of direction due to the lack of vision caused by fire during a flashover.

disorientation secondary to prolonged zero visibility conditions. The loss of direction due to the lack of vision caused by heavy smoke during prolonged zero visibility conditions.

disorientation secondary to wind-driven fire. The loss of direction due to the lack of vision caused by fire during a wind-driven fire.

double attack-supply evolution. A contemporary hose evolution in which the first two arriving engine companies attack the fire while the third establishes a supply line to be used by both.

dual pumping. An evolution in which two engines share water from a single hydrant.

dual pumping at a hydrant at the scene. A contemporary evolution in which two engines share water from a single hydrant while engines and hydrant are in close proximity to the fire scene.

dual pumping at the hydrant. A traditional evolution in which two engines share the water from a single hydrant while engines are in close proximity to the hydrant and at a distance from the fire scene.

dual pumping at the scene. A contemporary evolution in which two engines share the water from a single distant hydrant while engines are in close proximity to the fire scene.

enclosed space. An area within an opened or enclosed structure with an absence of openings for prompt ventilation (e.g., ceiling spaces, attic spaces, and high-rise hallways) or emergency evacuation (e.g., basements).

enclosed structure. A structure with an absence of windows or doors to provide prompt ventilation and emergency evacuation.

enclosed structure standard operating guideline. A preestablished flexible incident action plan to avoid life-threatening hazards that lead to firefighter disorientation during enclosed structure fires.

enclosed structure tactics. Tactics that combine a careful and calculated approach to interior firefighting at enclosed structure fires, by use of cautious interior assessments and, when possible, short interior attacks.

enclosed structure with a basement. A structure with an absence of windows or doors to provide prompt ventilation and emergency evacuation from either grade, above grade, or basement level.

extremely dangerous. Term used to describe the degree of danger associated with an enclosed structure fire in which traumatic firefighter fatalities occur at a disproportionate rate.

firefighter disorientation. The loss of direction due to the lack of vision in a structure fire.

firefighter disorientation sequence. A chain of events leading to firefighter disorientation including fire in an enclosed structure with smoke

Glossary

showing, a quick and aggressive interior attack, deteriorating conditions, handline separation, and firefighter disorientation.

heavy smoke. Dense smoke that blinds a fully bunkered firefighter.

high-rise hallway. An enclosed space within a high-rise structure having a rectangular and horizontal configuration.

high-rise stairwell. An enclosed space within a high-rise structure having a rectangular and vertical configuration.

hot smoke. Smoke of increased temperature that causes pain or injury to a fully bunkered firefighter.

initial size-up. The rapid collection and evaluation of fireground factors that determine the actions required to safely extinguish a structure fire.

large enclosed structure. Term used to generally define the size of a large enclosed structure by use of approximate dimensions of 100 ft. by 100 ft. or greater.

opened structure tactics. Tactics used during execution of an offensive strategy at opened structure fires that include a quick and aggressive interior attack from the unburned side of the structure.

opened structure with a basement. A structure with sufficient windows or doors to provide prompt ventilation and emergency evacuation from grade but not from basement level.

opened structure. A structure with sufficient windows or doors to provide prompt ventilation and emergency evacuation.

prolonged zero visibility conditions (PZVCs). Heavy smoke conditions that last longer than 15 minutes.

protected enclosed structure. A structure with an absence of windows or doors to provide prompt ventilation and emergency evacuation, which is fully protected by an operable sprinkler system.

protected high-rise structure. An enclosed multistory structure protected by an operable automatic sprinkler and standpipe system.

quick backup line. Within the safe extinguishing limitations of the water in a booster tank, a second handline that is used to back up the first as soon as the second engine arrives and while a third engine lays a supply line.

short interior attack. An enclosed structure tactic that minimizes the travel distance between the exterior and the seat of the fire to maximize firefighter safety.

single-engine evolution. A traditional hose evolution in which an engine initiates a straight or forward lay from a hydrant to establish its own supply line.

single-engine evolution with dual pumping at the scene. A contemporary hose evolution in which an engine initiates a straight or forward lay from a hydrant to establish its own supply line. Thereafter, the second engine makes a dual connection with the first engine to obtain a supply of unused water.

size-up factors. Fireground factors that collectively indicate the action that can be taken to safely extinguish a structure fire, including but not limited to wind speed and direction, structure type (opened or enclosed), type of occupancy, life safety hazard, type and amount of smoke showing, amount of fire showing, type of construction, use of trusses, use of dimensional lumber, age of the structure, condition of the structure, presence of basement windows or doors, presence of a slope or drop off along the foundation line, presence of boarded or barred windows, and number of hydrants available.

tenable smoke. Smoke that does not cause pain or injury to a fully bunkered firefighter.

unprotected enclosed structure. A structure with an absence of windows or doors to provide prompt ventilation and emergency evacuation, which is not protected by an operable sprinkler system.

unprotected high-rise structure. An enclosed multistory structure not protected by an operable automatic sprinkler or standpipe system.

wind-driven fire. An extremely dangerous structure fire in which a 10 mph or stronger wind speed pressurizes fire venting from a window, door, or roof, causing it to rapidly spread when an interior flow path is established.

working fire. A sizable amount of fire present during a structure fire that requires establishment of a supply line to fully extinguish.

zero visibility conditions. Heavy and blinding smoke conditions that last 15 minutes or less.

INDEX

A

accountability sector officer, 275, 286, 290
    in Worcester, Massachusetts, 301
aggressive interior attack
    backdraft and, 114, 118
    in basements, 62, 172, 236, 238–239, 243
    in Carthage, Missouri, 305, 307
    in Charleston, South Carolina, 324–325
    with double attack-supply evolution, 205, 214
    in enclosed spaces, 108
    in enclosed structures, 108
    fatalities in, 110–111
    firefighter disorientation and, 95
    flashover and, 125, 129, 133
    hose evolutions and, 183
    in Houston, Texas, 103
    in large structures, 3, 256–257, 259, 267, 268, 288–289
    in Memphis, Tennessee, 302
    in opened structures, 102
    in Phoenix, Arizona, 313, 314
    size-up and, 106–107
    standard operating procedures for, 277
    in wind-driven fires, 152, 161
    in Worcester, Massachusetts, 297
    worst-case scenario for, 27
Agra, India, 356
American Water Works Association (AWWA), 184
    on hydrants, 192–193
"Analysis of a Fatal Wind-Driven Fire in a Single-Story House" (Barowy and Madrzykowski), 142–143

"Analysis of Structural Firefighter Fatality Database" (Mora), 110
Annapolis, Maryland, two-story retail structure in, 13
apartment buildings
    as dangerous, 45
    as extremely dangerous, 44
    flashover in, 20
    in Maryland, 3
    sprinkler system in, 45
    as unprotected, 33, 44
area familiarity, for basements, 59
Arkansas City, Kansas, 164–166
Asheville, North Carolina, office building in, 3
Asunción, Paraguay, 356
attack pumpers
    attack-supply evolution and, 194
    in Charleston, South Carolina, 324–325, 331, 338
    dual pumping and, 189–190
    feeder hose on, 291
    hydrants at great distances and, 342
    in large structures, 291
    piston intake relief valves on, 196
attack-supply evolution (blitz attack). *See also* double attack-supply evolution
    attack pumpers and, 194
    in Charleston, South Carolina, 333–335
    dual pumping and, 197–207
    for hose evolutions, 184–186
    three-engine response and, 199–200
attic spaces
    backdraft in, 113, 118

wind-driven fires and, 164–166
automatic fire suppression systems. *See* sprinkler systems

B

backdraft, 113–124
    aggressive interior attack and, 114, 118
    in attic spaces, 113, 118
    avoidance of, 115
    in basements, 64, 113, 242
    in Boston, Massachusetts, 121–123
    in Brussels, Belgium, 350
    ceilings and, 350
    in Chicago, Illinois, 123–124
    in Colorado, 117–118
    in Crooksville, Ohio, 118–120
    defined, 87, 113
    doors and, 124
    double attack-supply evolution and, 214
    in enclosed structures, 27, 87, 124, 235
    evacuation from, 120
    firefighter disorientation with, 87, 122
    handlines and, 95–96, 120
    in Memphis, Tennessee, 120–121, 301, 305
    in opened structures, 27, 122
    in Pittsburgh, Pennsylvania, 114
    prolonged zero-visibility conditions and, 123
    in restaurants, 117–118
    roof collapse from, 88
    signs of, 116, 122–123
    size-up and, 107
    thermal imaging camera for, 121
    in two-story garden apartments, 20
    in vacant commercial structure, 123–124
    windows and, 124
backup company, for large structures, 282
backup lines
    for basements, 242
    in double attack-supply evolution, 211–213
    flashover and, 212
Bangkok, Thailand, 357
bar joists, 223
Barowy, A., 142–143
basements
    aggressive interior attack in, 62, 172, 236, 238–239, 243
    area familiarity for, 59
    backdraft in, 113
    burglar bars to, 61
    as dangerous, 62
    doors in, 50, 56, 58
    enclosed structures with, 59–61
    as extremely dangerous, 51, 62–63
    flashover in, 64, 240, 242
    floor collapse into, 63, 88, 96, 172, 237–238, 246, 248, 249, 250–251
    full-fog nozzles in, 241
    ground ladders in, 246, 253–254
    high-expansion foam in, 247
    inspection holes for, 59, 241
    occupancy type of, 243
    occupied, 248–255
    opened structures with, 50–59, 236–256
    piercing nozzles in, 246–247, 255–256, 353
    in Pittsburgh, Pennsylvania, 52, 62
    prefire surveys for, 58
    primary search of, 239–244, 248–255
    questioning occupants about, 58
    quick backup hoselines for, 242
    rapid intervention teams for, 254
    security bars for, 50, 61
    signs of, 54–59
    single-family structures with, 237
    situational awareness in, 62–64
    size-up for, 54–59, 238–244, 251
    stairways of, 242–243
    steps to, 57–58
    thermal imaging camera for, 162, 241

360-degree walk-around for, 54–55, 241
unoccupied, 244–248
as unprotected, 62, 236
vacant commercial structure with, 60–61
wind-driven fire in, 167, 240–241
windows in, 53, 56, 253
Baytown, Texas, wind-driven fires in, 152
Beslan, Russia, 356
"big box" structures
as protected, 67
with sprinkler system, 67
blitz attack. *See* attack-supply evolution
Blue Diamond Firefighter Hazard ID Marking System, 227
boarded doors and/or windows
in enclosed structures, 21
opened structures and, 12
Bolin, Texas, Maxim Egg Farm in, 5
booster tanks, 185
for basements, 245
in Charleston, South Carolina, 325, 330
defensive attacks and, 187–188
in double attack-supply evolution, 211, 212–213
Boston, Massachusetts, backdraft in, 121–123
Bowker, Gary, 227
bowstring trusses, 225
breaching
in Charleston, South Carolina, 332, 335–336
firefighter disorientation and, 353
in large structures, 269–272
Phoenix, Arizona and, 317–318
for short interior attack, 175
for wind-driven fires, 147
breathing difficulty, in wind-driven fires, 153–154
brick and mortar over windows, in enclosed structures, 21, 22, 40
Brussels, Belgium, 350
Bryan, Texas, 4
Bryner, Nelson, 343
Buenos Aires, Argentina, 356

Bureau of Alcohol, Tobacco, Firearms and Explosives (ATF), 321
burglar bars
to basements, 61
on enclosed structures, 21, 22, 23, 24
on large structures, 268
opened structures and, 12
on residential structures, 43

C

Caracas, Venezuela, 356
Carthage, Missouri, wind-driven fire in, 305–309
Casablanca, Morocco, 357
cautious interior assessment
for large structures, 258–267, 281
standard operating guidelines for, 307–308
ceiling spaces
backdraft in, 113, 120, 122
in Carthage, Missouri, 306
in Charleston, South Carolina, 326, 336, 337
in fire through roof, 345
ground ladders and, 121
in large structures, 261
in Memphis, Tennessee, 301
in Phoenix, Arizona, 315
thermal imaging camera for, 222, 274
trusses in, in attack spaces, 222
in wind-driven fires, 164, 309
ceilings. *See also* flashover
backdraft and, 350
straight stream nozzles for, 137, 232
as wind control devices, 146–147
cerebrovascular accidents (CVAs), 110
Charleston, South Carolina
attack pumpers in, 324–325, 331, 338
commercial structures in, 102, 174, 320–340
enclosed structure in, 102, 174, 320–340
entanglement hazards in, 325–326

evacuation in, 323
firefighter disorientation in, 102
hydrants in, 340–341
prolonged zero-visibility conditions in, 102, 174, 320–340
roof collapse in, 326, 344
unprotected at, 320, 328
vertical ventilation in, 343–344
wind-driven fire in, 329
Chicago, Illinois
    backdraft in, 123–124
    roof collapses in, 3
    wind-driven fires in, 152
Chittagong, Bangladesh, 356
Cincinnati, Ohio, flashover in, 86
cinder block over windows
    in enclosed structures, 21, 24
    in Fall River, Massachusetts, 39
    on five-story structure, 35
Clement, Clifford, 97
Cold Storage Warehouse, in Worcester, Massachusetts, 83, 102, 296–301
collapse. *See* floor collapse; roof collapse
collapse zones
    in Charleston, South Carolina, 338
    defensive attack and, 106, 265, 301, 320, 338
    in Phoenix, Arizona, 320
Colorado, backdraft in, 117–118
commercial structures. *See also* grocery stores; high-rise structures; restaurants; warehouses
    backdraft in, 123–124
    with basement, 60–61
    in Charleston, South Carolina, 102, 174, 320–340
    degree of danger in, 43–44
    as enclosed structure, 22
    in Memphis, Tennessee, 301–305
    as opened structure, 43
    as protected, 44, 91
    vacant, 60–61, 123–124
company integrity
    firefighter disorientation and, 96, 352
    in large structures, 285–287
    in wind-driven fires, 168

construction type
    of enclosed structures, 47–48
    firefighter disorientation and, 93, 221–228, 352
    floor collapse and, 222–228
    floor collapse into basements and, 64
    of large structures, 70
    protected and, 48
    roof collapse and, 222–228
    size-up for, 228
    vertical ventilation and, 231
conversion steam
    firefighter disorientation with, 91
    full-fog nozzles and, 137
    from knock down, 274
    in Phoenix, Arizona, 317
    straight stream nozzles and, 137
    ventilation for, 137
Crooksville, Ohio, backdraft in, 118–120

D

danger, degrees of. *See also* extremely dangerous
    in commercial structures, 43–44
    for enclosed structures, 27–28
    fall height criteria and, 32–34
    for firefighter disorientation, 25–35
    in large structures, 28–29, 69–76
    occupancy type and, 43–44
    in opened structures, 26–27
    in opened structures with basement, 51
    in residential districts, 43–44
    sprinkler systems and, 40, 44, 45
    in tall structures, 30–31
    unprotected and, 33
dangerousness
    of apartment buildings, 45
    of basements, 62
    of high-rise structures, 45
    of large structures, 69–70, 71, 72, 73, 74
    of opened structures, 26–27, 43
    of shopping malls, 42

sprinkler systems and, 40–41
of three-story structures, 45
defensive attack
    booster tanks and, 187–188
    in Charleston, South Carolina, 338
    collapse zones and, 106, 265, 301, 320, 338
    construction type and, 228
    for large structures, 258, 260, 274–275, 282
    in Oak Park, Michigan, 154
    in Phoenix, Arizona, 313
    roof collapse and, 258
    size-up and, 106
    unprotected and, 258
detention centers, 74
Deutsche Bank, in New York City, 35
Dhaka, Bangladesh, 357
directional arrows
    in high-rise structures, 169
    in large structures, 266
    in Phoenix, Arizona, 319
    for prolonged zero-visibility conditions, 175
disorientation. *See* firefighter disorientation
dispatchers
    extremely dangerous and, 176
    primary search and, 240, 249
    residual pressure and, 176
    unprotected and, 319
    wind-driven fire and, 152, 159, 163, 308, 315–316
    wooden trusses and, 308
doors. *See also* boarded doors and/or windows
    backdraft and, 124
    in basements, 50, 56, 58
    in Charleston, South Carolina, 324, 335, 337
    enclosed structures and, 21–25
    for evacuation, 14
    fire dynamics and, 132
    flow path and, 132
    in high-rise structures, 169
    in large structures, 268
    in opened structures, 12–20
    in Phoenix, Arizona, 317
    for ventilation, 15
    as wind control devices, 146–147
    in wind-driven fires, 144–146, 309
    in working fires, 304
double attack-supply evolution
    aggressive interior attack with, 205
    backdraft and, 214
    in basements, 236, 245, 250
    benefits of, 211–217
    in Charleston, South Carolina, 330, 332–338
    dual pumping and, 200–204
    in enclosed structures, 317
    in fire through roof, 346
    flashover and, 214
    friction loss in, 204
    in large structures, 287, 288, 289, 291
    Memphis, Tennessee and, 303–304
    portable interior monitors in, 214–216
    quick backup hoselines in, 211–213
dual pumping, 188–204
    attack pumpers and, 189–190
    attack-supply evolution and, 197–207
    for basements, 245
    in Charleston, South Carolina, 332–338
    double attack-supply evolution and, 200–204
    with 5-inch hose, 190–195
    for large structures, 291
    single-engine evolution and, 195–197
    three-engine response and, 199–200
    with 2-1/2 or 3-inch supply hose, 189–190
duplexes
    basements in, 53
    rollover in, 170–171
    vacant, 170–171

E

Ebenezer Baptist Church, in Pittsburgh, Pennsylvania, 62, 292
egress. *See* evacuation
El Jadida, Morocco, 356
electric utility disconnect, 284, 290
electrical fire, in Pittsburgh, Pennsylvania, 114
electrocution, 110
    in two-story garden apartments, 20
enclosed spaces, 65–67. *See also* basements; ceiling spaces; large structures
    aggressive interior attack in, 108
    firefighter disorientation and, 352
    piercing nozzles in, 352
    in residential structures, 51
    subway stations and tunnels as, 42
enclosed structures, 21–25
    age of, 48
    aggressive interior attack in, 108
    backdraft in, 27, 87, 113, 124, 235
    with basements, 59–61
    in Carthage, Missouri, 305–309
    in Charleston, South Carolina, 102, 174, 320–340
    configuration of, 48
    construction type of, 47–48
    conversion of, 47
    defined, 21
    degree of danger for, 27–28
    double attack-supply evolution for, 317
    as extremely dangerous, 26, 37, 41
    in Fall River, Massachusetts, 39–40
    fire through roof in, 345–347
    firefighter disorientation in, 94, 235–347
    flashover in, 27, 235
    floor collapse in, 27, 235
    four-story commercial structure as, 22
    fuel loading in, 27
    grocery stores as, 40
    handlines in, 103–104
    Incident Command System for, 278–279
    large structures as, 28–29, 67–76, 75
    in Memphis, Tennessee, 301–305
    occupancy types for, 47, 293–294
    opened structures with, 64–65
    in Phoenix, Arizona, 157, 232, 310–320
    preplanning for, 175–177
    primary search in, 293–294
    as protected, 35–36, 42
    residential structures as, 28, 43–44
    shopping malls as, 42
    size of, 48
    size-up for, 104, 107
    sprinkler systems and, 38, 40, 175–177
    standard operating guidelines for, 235, 275, 276, 278–279, 280
    standpipes in, 38
    subway stations and tunnels as, 42
    tactics and guidelines for, 235–294
    tall structures as, 30–31
    three-story structures as, 44
    as unprotected, 4, 5, 21, 23, 35–36, 176, 289, 338, 349–350, 355–357
    ventilation for, 230
    vertical ventilation in, 232, 343–344
    warehouses as, 40
    wind-driven fire in, 344
    in Worcester, Massachusetts, 296–301
    worst-case scenario for, 27–28
entanglement hazards, 64, 95, 100
    in Boston, Massachusetts, 123
    in Carthage, Missouri, 309
    in Charleston, South Carolina, 325–326
    in Fall River, Massachusetts, 39
    in fire through roof, 346
    in Goldsboro, North Carolina, 68
    in Houston, Texas, 103
    prolonged zero-visibility conditions and, 81, 174
    in Worcester, Massachusetts, 300
Erwadi, India, 356

evacuation
- from backdraft, 120
- burglar bars and, 24
- in Carthage, Missouri, 307
- in Charleston, South Carolina, 323, 324
- doors for, 14
- enclosed structures and, 21–25
- firefighter disorientation and, 352, 353
- from flashover, 85, 127
- hose evolutions and, 183
- from large structures, 28–29, 265–266
- in Memphis, Tennessee, 301
- from opened structures, 19–20
- prolonged zero-visibility conditions and, 319
- short interior attack and, 273
- windows for, 14, 16–17

Evaluating Fire Fighting Tactics Under Wind Driven Conditions (Kerber and Madrzykowski), 142

Everyone Goes Home Firefighter Life Safety Initiatives Program (FLSI), 2

extremely dangerous
- apartment buildings as, 44
- basements as, 51, 62–63
- dispatchers and, 176
- enclosed structures as, 26, 37, 41
- grocery stores as, 75
- in high-rise stairwells, 66
- high-rise structures as, 44
- large structures as, 67, 70, 73, 75
- residential structures as, 44
- three-story structures as, 33, 34
- unprotected and, 21, 23, 28, 29
- vacant two-story structure as, 75, 76
- in Worcester, Massachusetts, 299–300

F

Fahy, Rita F., 2

fall height criteria
- degree of danger and, 32–34
- unprotected and, 33

Fall River, Massachusetts, 39–40, 97

fast and aggressive interior attack, 286

feeder hose
- on attack pumpers, 291
- in large structures, 261

FFIPP, "U.S. Firefighter Disorientation Study 1979–2001" and, 6

fire blankets, 140–142
- ground ladders for, 141

fire curtains, 140–142
- in Charleston, South Carolina, 332
- ground ladders for, 162
- for prolonged zero-visibility conditions, 175
- for wind-driven fires, 161

fire dynamics, 131–132. *See also* flow path

"Fire Dynamics: The Science of Fire Fighting" (Madrzykowski), 133

fire escapes, blocking of, 35

fire through roof, in enclosed structures, 345–347

firefighter disorientation
- aggressive interior attack and, 95
- with backdraft, 87, 122
- in Charleston, South Carolina, 102
- company integrity and, 96
- construction type and, 93, 221–228
- with conversion steam, 91
- defined, 8
- degree of danger for, 25–35
- in enclosed structures, 94, 235–347
- fatality problem with, 1–5
- with flashover, 84–87, 125, 126–128
- with floor collapse, 88–89
- handlines and, 95–96
- in high-rise structures, 168–170
- hose evolutions and, 181–220
- in large structures, 288–289
- occupancy type and, 93

prevention of, 350–352
with prolonged zero-visibility
  conditions, 82–84, 95, 102
with roof collapse, 88–89
sequence of, 99–104
similarities with, 93–97
size-up and, 105–111, 351
softening techniques and, 352
sprinkler systems and, 97
structure age and, 94
structure size and, 93
types of, 82–91
"U.S. Firefighter Disorientation
  Study 1979–2001" on, 6–8, 93,
  108
ventilation and, 229–232
with wind-driven fire, 89–90
in working fires, 11
"Firefighter Fatalities in the United
  States—2024" (Fahy, LeBlanc, and
  Molis), 2
Fire Fighter Fatality Investigation and
  Prevention Program (FFFIPP), 2,
  296, 310
  on Phoenix, Arizona, 310
  on quick backup hoseline, 211
Fire Fighter Fatality Investigation and
  Prevention Program, of National
  Institute for Occupational Safety
  and Health, 118
Firefighter Safety Research Institute
  (FSRI), 131
First Interstate Tower, in Los Angeles,
  37–38
five-story structure
  cinder block over windows on, 35
  plywood sheeting over windows on,
  35
flameover. *See* rollover
flashover, 124–138
  aggressive interior attack and, 125,
  129, 133
  avoidance of, 126–128
  in basements, 64, 240, 242
  in Cincinnati, Ohio, 86
  defined, 84, 124
  double attack-supply evolution and,
  214

in enclosed structures, 27, 235
evacuation from, 85
firefighter disorientation with,
  84–87, 125, 126–128
handlines and, 95–96
hose evolutions and, 183
in opened structures, 27
in Pensacola, Florida, 86
protected and, 130–131
quick backup hoselines and, 212
rollover and, 170–171, 260–261
signs of, 126–128, 130, 133–134
size-up and, 106, 107
softening techniques for, 134–139
sprinkler system for, 130–131
in two-story garden apartments, 20
unprotected and, 131, 249
ventilation and, 133–134, 230
in Worcester, Massachusetts, 300
zero-visibility and, 130
flashover contingency plan, 128–131
full-fog nozzles for, 140
for self-venting, 140
floor collapse, 172–173
  into basement, 63, 88, 96, 172,
  237–238, 246, 248, 249, 250–251
  construction type and, 222–228
  in enclosed structures, 27, 235
  firefighter disorientation with,
  88–89
  handlines and, 96
  joists and, 224
  in opened structures, 27
  personal protective equipment in,
  246
  size-up and, 107
  in two-story garden apartments, 20
  wind-driven fire and, 344
  in Worcester, Massachusetts, 300
flow path
  in fire dynamics, 132–133
  in large structures, 264
  in wind-driven fires, 133, 146, 150,
  159, 164
four-story structures
  as enclosed structures, 22
  fall height criteria and, 33
  vacant, 24

Index

friction loss, 192, 193
   in double attack-supply evolution, 204
   hydrants at great distances and, 342, 343
fuel loading
   in enclosed structures, 27
   knock down and, 27
   in large structures, 70
   in opened structures, 27
   vertical ventilation and, 343
   working fires with, 27
full-fog nozzles
   in basements, 241
   in Charleston, South Carolina, 336
   conversion steam and, 137
   for flashover contingency plan, 140
   in large structures, 263

G

garden homes, basements in, 53
gas utility disconnect, 284, 290
Goldsboro, North Carolina, large structures in, 68
grocery stores
   as enclosed structures, 40
   as extremely dangerous, 75
   in Phoenix, Arizona, 157, 232, 310–320
   vertical ventilation in, 232
   wind-driven fire in, 157
ground ladders
   in basements, 246, 253–254
   ceiling spaces and, 121
   for fire blankets, 141
   for fire curtains, 162
   fire curtains from, 162
"Guidelines for Field Triage of Injured Patients: Recommendations of the National Expert Panel on Field Triage" (Sasser, Hunt, Sullivent), 32
gypsum board, in enclosed structures, 21

H

hallways
   in high-rise structures, 11, 66, 97, 103
   wind-driven fire and, 103, 145, 148, 168
handlines. *See also* entanglement hazards
   attack-supply evolution and, 184–186
   backdraft and, 95–96, 120
   in basements, 241, 245, 246
   in Charleston, South Carolina, 324–325, 330, 336, 337
   dual pumping to, 194
   in enclosed structures, 103–104
   in Fall River, Massachusetts, 39
   in fire through roof, 345–346
   firefighter disorientation and, 95–96, 352
   flashover and, 95–96, 128
   floor collapse and, 96
   hose evolutions and, 182, 183, 184–186
   hydraulic ventilation and, 229
   in large structures, 68, 258, 261, 281, 291
   in Memphis, Tennessee, 301, 303
   in opened structures, 103
   in Phoenix, Arizona, 313
   in primary search, 95
   in prolonged zero-visibility conditions, 83, 103
   in wind-driven fires, 168, 169
   in Worcester, Massachusetts, 297
heart attacks, 110
Hermosillo, Mexico, 357
high-expansion foam, in basements, 247
high-rise structures
   as dangerous, 45
   as extremely dangerous, 44
   firefighter disorientation in, 168–170
   First Interstate Tower in Los Angeles, 37–38
   hallways in, 11, 66, 97, 103
   in Houston, Texas, 103

One Meridian Plaza in Philadelphia, 35–36
  as protected, 38
  size-up for, 143–144
  sprinkler systems and, 36, 38, 45
  stairwells in, 66–67, 90
  standard operating guidelines for, 168, 300
  as unprotected, 36, 44
  wind-driven fires in, 143–152, 168–170
hills, basements on, 56–57
horizontal ventilation, 229
hose evolutions
  aggressive interior attack and, 183
  attack-supply evolution for, 184–186
  contemporary approaches to, 188–207
  dual pumping, 188–204
  firefighter disorientation and, 181–220, 352
  flashover and, 183
  greatest level of effectiveness of, 216
  hydrants and, 217–220
  single-engine evolution in, 186–188, 333–335
  traditional approaches to, 184–188
hotels, as high-rise structures, 45
Houston, Texas
  aggressive interior attack in, 103
  high-rise structures in, 103
  roof collapse in, 4, 88–89
  wind-driven fire in, 103, 152, 168
Hunt, R. C., 32
hydrants
  American Water Works Association on, 192–193
  in Charleston, South Carolina, 340–341
  distribution and spacing of, 184
  for dual pumping, 192–193
  flowcharts for fire flow determination for, 217–220
  at great distance from fire, 340–343
  hose evolutions and, 217–220
  prefire surveys for, 217
  residual pressure of, 217–220
  static pressure of, 217–220
hydrants at great distances, friction loss and, 342, 343
hydraulic ventilation, 230

I

Icheon, South Korea, 357
I-joists, 226–227
immediately-dangerous-to-life-and-health (IDLH), 25
Incident Command System (ICS), 26
  for enclosed structures, 278–279
initial size-up. *See* size-up
inspection holes, for basements, 59, 241
interior monitors. *See* portable interior monitors
International Association of Fire Chiefs (IAFC)
  on backdrafts, 115
  on Charleston, South Carolina, 327–328
  Safety Health and Survival Section of, 2
International Association of Fire Fighters (IAFF), 2
International Building Code, 221
International Fire Service Journal of Leadership and Management, 133
International Society of Fire Service Instructors (ISFSI), 136

J

Jilin, China, 356
joists
  bar, 223
  floor collapse and, 224
  I-joists, 226–227

Index

K

Kamyshevatskaya, Russia, 356
Kerber, Stephen, 89, 137–138, 142–143
   on door control, 144–145
knock down, 127
   in basements, 250–251, 253
   conversion steam from, 274
   in fire through roof, 346
   fuel loading and, 27
   hose evolutions and, 182
   in large structures, 274
   master stream for, 169
   in Memphis, Tennessee, 301, 304
   portable interior monitors for, 28
   in wind-driven fires, 161

L

large structures
   age of, 70
   aggressive interior attack in, 3, 256–257, 259, 267, 268, 288–289
   attack pumpers in, 291
   backup company for, 282
   breaching of, 269–272
   cautious interior assessment for, 258–267, 281
   company integrity in, 285–287
   conversion steam in, 91
   as dangerous, 69–70, 71, 72, 73, 74
   defensive attack for, 258, 260, 274–275, 282
   degree of danger in, 28–29, 69–76
   double attack-supply evolution for, 287, 288, 289
   double attack-supply evolution in, 291
   as enclosed structures, 67–76, 75
   evacuation from, 265–266
   as extremely dangerous, 67, 70, 73, 75
   in Fall River, Massachusetts, 39–40, 97
   firefighter disorientation in, 288–289
   flow path in, 264
   fuel loading in, 70
   full-fog nozzles in, 263
   in Goldsboro, North Carolina, 68
   handlines in, 68, 258, 261, 281, 291
   knock down in, 274
   leader lines in, 261, 291
   maximizing flow in, 290–292
   occupancy type of, 281
   portable interior monitors in, 261–262, 282, 289
   as protected, 70, 71, 72, 74
   rapid intervention teams in, 258, 268, 283
   in residential districts, 29
   rollover in, 260–261
   runoff from, 292–293
   short interior attack in, 268–274, 282, 284, 287
   size-up for, 68, 257–258, 281
   sprinkler system in, 70, 74
   standard operating guidelines for, 275, 280
   tall structures as, 30
   thermal imaging camera for, 257, 258–259, 260, 268
   360-degree walk-around for, 257, 258, 289
   truck company in, 290
   unknown-known guidelines for, 256–292
   as unprotected, 5, 29, 67, 68, 70, 75, 114, 289, 338
   water damage from, 292–293
   zero-visibility in, 69, 260, 265–266
leader lines, in large structures, 261, 291
LeBlanc, Paul R., 2
Lima, Peru, 356
Los Angeles, First Interstate Tower in, 37–38

M

Machakos, Kenya, 356
Madison Square Garden, 48

Madrzykowski, Daniel, 89
    on door control, 144–145
    on fire dynamics, 131–132
    on flow path, 132–133
    wind-driven fires, 139
    on wind-driven fires, 142–143
Mary Pang fire, in Seattle, Washington, 62
Maryland, apartment building in, 3
master stream, 6
    in Boston, Massachusetts, 123
    degree of danger and, 25
    in double attack-supply evolution, 209
    in dual pumping, 192
    in fire through roof, 345–346
    for knock down, 169
    in single-engine evolution, 169–186
    in wind-driven fires, 147
Maxim Egg Farm, in Bolin, Texas, 5
Mayday incident
    Incident Command System for, 279
    in large structures, 268
    in Phoenix, Arizona, 313
    short interior attack and, 272
    in Toledo, Ohio, 4
Medan, Indonesia, 357
Memphis, Tennessee
    backdraft in, 120–121, 301, 305
    commercial structures in, 301–305
    enclosed structure in, 301–305
    knock down in, 301, 304
    size-up in, 301
metal sheeting, in enclosed structures, 21
metal-framed windows, 17
mobile homes
    in West Virginia, 153
    wind-driven fire in, 152
Molis, Joseph L., 2
Molo, Kenya, 357
Mora, W. R., 6–8, 93, 108, 110, 256
Moscow, Russia, 356
multifamily homes. *See also* apartment buildings
    basements in, 53
Muncie, Illinois, roof collapses in, 3
museums, as enclosed structures, 25

N

National Fallen Firefighters Foundation (NFFF), 310
    on backdrafts, 115
    Everyone Goes Home Firefighter Life Safety Initiatives Program of, 2
National Fire Academy, 325
National Fire Protection Association (NFPA), 2
    on backdraft, 113
    on flashover, 85
    NFPA 555: Guide on Methods for Evaluating Potential for Room Flashover, 113
    on quick backup hoseline, 211
    on residual pressure, 208
National Institute for Occupational Safety and Health (NIOSH)
    on Charleston, South Carolina, 321, 343
    Firefighter Fatality Investigation and Prevention Program of, 2, 118, 211, 296, 310
    on I-joists, 226–227
    on opened structures, 277
    on precooling, 232
    "U.S. Firefighter Disorientation Study 1979–2001" and, 6
    on wind-driven fire, 89–90
National Institute of Standards and Technology (NIST), 2, 134–137
    on high-rise structures, 143
    on self-venting, 140
    on wind control devices, 141
    on wind speed, 156–157
    on wind-driven fires, 148–151, 163
natural gas explosion, in two-story garden apartments, 20
New York City
    Deutsche Bank in, 35
    Madison Square Garden in, 48
    wind-driven fires in, 152, 168
New York City Fire Department (FDNY), 333
    at two-story structure, 134

on wind-driven fires, 148–150
NFPA 555: Guide on Methods for Evaluating Potential for Room Flashover, National Fire Protection Association, 113

O

Oak Park, Michigan, wind-driven fire in, 154–155, 158
occupancy types
    of basements, 243
    degree of danger and, 43–44
    of enclosed structures, 47, 293–294
    firefighter disorientation and, 93
    of large structures, 281
Occupational Safety and Health Administration, two-in/two-out rule of, 184
Offensive Strategy. *See* aggressive interior attack
office building, in Asheville, North Carolina, 3
offices, as enclosed structures, 25
One Meridian Plaza, in Philadelphia, 35–36
opened structures
    aggressive interior attack in, 102
    backdraft in, 27, 122
    with basements, 50–59, 236–256
    commercial structures as, 43
    as dangerous, 26–27, 43
    defined, 12–20
    degree of danger in, 26–27
    double attack-supply evolution in, 214
    with enclosed structures, 64–65
    evacuation from, 19–20
    fall height criteria for, 33
    flashover in, 27
    floor collapse in, 27
    fuel loading in, 27
    handlines in, 103
    in residential districts, 18–20
    residential structures as, 277
    single-family structures as, 12
    size-up for, 102
    standard operating guidelines for, 288
    two-story garden apartment as, 20
    two-story structures as, 15
    ventilation in, 19
    working fires in, 54
    worst-case scenario for, 26–27
Our Lady of Angels Fire Memorial, 355
oxygen, backdraft and, 113–114, 120, 121

P

Palembang, Indonesia, 356
Pensacola, Florida, 12
    flashover in, 86
percentage method, 217–220
Perm, Russia, 357
personal protective equipment (PPE)
    flashover and, 126
    in floor collapse, 246
    in primary search, 249
    "U.S. Firefighter Disorientation Study 1979–2001" and, 6
personnel accountability report (PAR), 275, 284, 290, 301
    in Carthage, Missouri, 307
    in Charleston, South Carolina, 338
    in Phoenix, Arizona, 320
Philadelphia, Pennsylvania, One Meridian Plaza in, 35–36
Phoenix, Arizona
    breaching and, 317–318
    enclosed structure in, 157, 232, 310–320
    grocery store in, 157, 232, 310–320
    plywood sheeting over windows in, 312
    Southwest Supermarket in, 157, 232, 310–320
    unprotected at, 315
    wind-driven fire in, 157, 232, 310–320, 313, 314
photoluminescence technology
    firefighter disorientation and, 353
    in high-rise structures, 169

Pictometry International Corporation, 321–322
piercing nozzles
  in basements, 246–247, 255–256, 353
  in enclosed spaces, 352
  firefighter disorientation and, 353
piston intake relief valves (PIRVs), 190–191, 195–198
  on attack pumpers, 196
  in double attack-supply evolution, 203
Pittsburgh, Pennsylvania
  backdraft in, 114
  basement in, 52, 62
  Ebenezer Baptist Church in, 62, 292
  electrical fire in, 114
plywood sheeting, over windows and/or doors
  to basement, 61
  in enclosed structures, 21, 22
  on five-story structure, 35
  in Phoenix, Arizona, 312
  on residential structures, 44
portable interior monitors
  in Charleston, South Carolina, 330, 331, 335, 336
  in double attack-supply evolution, 214–216
  for knock down, 28
  in large structures, 261–262, 282, 289, 292
  in Phoenix, Arizona, 317, 319–320
positive pressure ventilators (PPVs), 147
  in basements, 245, 253
  in high-rise structures, 169
  horizontal ventilation and, 229
  Phoenix, Arizona and, 317–318
  property damage and, 163
  in wind-driven fires, 163
precooling
  firefighter disorientation and, 352
  at residential structures, 232
prefire surveys
  for basements, 58
  for hydrants, 217
primary search, 3

of basements, 239–244, 248–255
dispatchers and, 240, 249
in enclosed structures, 293–294
floor collapse and, 172
handlines in, 95
hose evolutions and, 182, 183
in Houston, Texas, 4
Memphis, Tennessee and, 304
prolonged zero-visibility conditions and, 82–83
vacant warehouses and, 301
in wind-driven fires, 161, 168
in Worcester, Massachusetts, 297
Prince George's County, Maryland, 167
Prince William County, Virginia, 167
  wind-driven fires in, 152
prolonged zero-visibility conditions (PZVCs), 80–81, 173–175
  backdraft and, 123
  in basements, 242
  in Carthage, Missouri, 309
  in Charleston, South Carolina, 102, 174, 320–340
  in enclosed structures, 235
  evacuation and, 319
  fire curtains for, 175
  in fire through roof, 346
  firefighter disorientation with, 82–84, 95, 102
  handlines in, 83, 103
  in high-rise structure fires, 168
  hose evolutions and, 182
  Memphis, Tennessee and, 305
  rapid intervention teams for, 175
  for short interior attack, 272
  in Worcester, Massachusetts, 297, 300
property damage. See also water damage
  in double attack-supply evolution, 202
  firefighter disorientation and, 353
  positive pressure ventilator and, 163
protected
  "big box" structures as, 67
  commercial structures as, 44, 91
  construction type and, 48
  enclosed structures as, 35–36, 42

in Fall River, Massachusetts, 39, 40
flashover and, 130–131
high-rise structures as, 38
large structures as, 70, 71, 72, 74
residential structures as, 44
shopping malls as, 42
tall structures as, 41

Q

Quezon City, Philippines, 356
quick backup hoselines
    for basements, 242
    in double attack-supply evolution, 211–213
    flashover and, 212

R

rapid intervention teams (RITs), 101, 103
    for basements, 254
    in Carthage, Missouri, 307
    firefighter disorientation and, 353
    in large structures, 258, 268, 283
    in Memphis, Tennessee, 301
    in Phoenix, Arizona, 313
    for prolonged zero-visibility conditions, 175
    for short interior attack, 272
    in Worcester, Massachusetts, 301
residential districts
    degree of danger in, 43–44
    large structures in, 29
    opened structures in, 18–20
    single-family structures in, 18
    two-story residential structure in, 18, 19
residential structures. *See also* apartment buildings; duplexes; mobile homes; multifamily homes; single-family structures
    enclosed spaces in, 51
    as enclosed structures, 28, 43–44
    as extremely dangerous, 44

as opened structures, 277
precooling at, 232
as protected, 44
as unprotected, 12, 28
wind-driven fires in, 152
residual pressure
    dispatchers and, 176
    in double attack-supply evolution, 208–210
    of hydrants, 217–220
    from hydrants at great distances, 342
restaurants
    backdraft in, 117–118
    in Carthage, Missouri, 305–309
    as enclosed structures, 25
reverse lays, in attack-supply evolution
    dual-pumping, 197–199
rollover
    flashover and, 170–171, 260–261
    in large structures, 260–261
roof collapse, 2, 3–4
    from backdraft, 88
    in Carthage, Missouri, 307
    in Charleston, South Carolina, 326, 336, 344
    in Chicago, 3
    construction type and, 222–228
    double attack-supply evolution and, 214
    in enclosed structures, 23, 27, 235
    firefighter disorientation with, 88–89
    in Houston, Texas, 4, 88–89
    in large spaces, 264–265
    Memphis, Tennessee and, 305
    in Muncie, Illinois, 3
    in opened structures, 27
    size-up and, 107
    before size-up and, 106
    in two-story garden apartments, 20
    in Wayne Westland, Illinois, 4
roof-access hatch
    in high-rise structures, 67, 147
    in wind-driven fires, 147
round configuration, of Madison Square Garden in New York City, 48

Routley, J. G., 81
runoff, from large structures, 292–293

S

Salisbury, North Carolina, warehouses in, 62, 344
San Antonio, Texas
    Hydrant Performance Chart of, 219
    prolonged zero-visibility conditions in, 81
Santiago, Chile, 357
Sasser, S. M., 32
scissor trusses, 225
Seattle, Washington, Mary Pang fire in, 62
security bars
    for basement doors, 50, 61
    in enclosed structures, 23
    in large structures, 268
self-contained breathing apparatus (SCBA), 29
    breathing air volume in, 81
    in Charleston, South Carolina, 326
    flashover and, 126
    in high-rise structure, 36
    in prolonged zero-visibility conditions, 81, 173–175
    in zero-visibility, 80
self-venting, 139–153
    flashover contingency plan for, 140
semifog nozzles
    in basements, 241
    in Charleston, South Carolina, 336
    in large structures, 263
    in Phoenix, Arizona, 320
Shanghai, China, 357
shopping malls, as enclosed structures, 42
short interior attack
    breaching and, 175
    in Charleston, South Carolina, 331, 335
    in large structures, 268–274, 282, 284, 287
    rapid intervention teams for, 272

single-engine evolution
    in Charleston, South Carolina, 333–335
    dual pumping and, 195–197
    in hose evolutions, 186–188, 333–335
single-family structures
    basements in, 53, 237
    as opened structure, 12
    in residential districts, 18
situational awareness, in basements, 62–64
size-up. *See also* 360-degree walk-around
    aggressive interior attack and, 106–107
    backdraft and, 107
    for basements, 54–59, 238–244, 251
    for construction type, 228
    defensive attack and, 106
    for enclosed structures, 104, 107
    firefighter disorientation and, 105–111, 351
    flashover and, 106, 107, 129
    floor collapse and, 107
    at large structures, 68
    for large structures, 257–258, 281
    in Memphis, Tennessee, 301, 303
    misreading factors of, 106–108
    for opened structures, 102
    speed of, 105–107
    for wind-driven fires, 146–147, 152, 308, 315
    Worcester, Massachusetts and, 299–300
smoke plume angle, in wind-driven fires, 156–158
softening techniques
    firefighter disorientation and, 352
    for flashover, 134–139
Solomon Guggenheim Museum in New York City, spiral configuration, 48
Southwest Supermarket, in Phoenix, Arizona, 157, 232, 310–320
spiral configuration, Solomon Guggenheim Museum in New York City, 48
sprinkler systems, 3. *See also* protected;

unprotected
  in apartment buildings, 45
  "big box" structures with, 67
  clogging of, 70
  degree of danger and, 40, 44, 45
  Deutsche Bank and, 35
  enclosed structures and, 21, 22, 38, 40, 175–177
  extremely dangerous and, 37, 41
  in Fall River, Massachusetts, 39
  firefighter disorientation and, 97
  at First Interstate Tower in Los Angeles, 37–38
  for flashover, 130–131
  high-rise structures and, 36, 38, 45
  large structures and, 28, 29, 70, 74
  at One Meridian Plaza in Philadelphia, 35–36
  opened structures and, 12, 27
  in shopping malls, 42
  three-story buildings and, 33
  Worcester, Massachusetts and, 300
St. Louis, Missouri, wind-driven fires in, 168
stairways
  backdraft in, 114
  of basements, 57–58, 242–243
  as enclosed spaces, 65
stairwells
  backdraft in, 113
  in high-rise structures, 66–67, 90, 147, 169
  in wind-driven fires, 147
  in Worcester, Massachusetts, 301
standard operating guidelines (SOG)
  for backdraft, 123, 124
  for basements, 236, 244, 246, 247, 250
  for cautious interior assessment, 307–308
  for dual pumping, 202–203
  for enclosed structures, 235, 275, 276, 278–279, 280
  for high-rise structures, 168, 300
  for hose evolutions, 182
  for large structures, 275, 280
  for opened structures, 288
  for short interior attack, 272
  for 360-degree walk-around, 307–308
  unknown-known guidelines and, 275–292
  for wind-driven fires, 146, 147, 163
standard operating procedures (SOP)
  for aggressive interior attack, 277
  backdraft and, 118
  in basements, 241
  for large structures, 286
  for size-up, 106–107
standpipes. *See also* protected; unprotected
  at Deutsche Bank, 35
  in enclosed structures, 38
  extremely dangerous and, 41
  fall height criteria and, 33
  in high-rise structures, 169
  Worcester, Massachusetts and, 300
static pressure, of hydrants, 217–220
steel trusses, 223
  in Charleston, South Carolina, 326
  in large spaces, 264
steps, to basement, 57–58, 242–243
Storz-to-Storz couplings, 196
straight lays, in attack-supply evolution dual-pumping, 197–199
straight stream nozzles
  for ceilings, 137, 232
  in Charleston, South Carolina, 336
  conversion steam and, 137
  in wind-driven fires, 149
subway stations and tunnels, as enclosed spaces, 42
Sullivent, E. E., 32

T

Taldykorgan, Kazakhstan, 357
tall structures. *See also* four-story structures; high-rise structures; three-story structures
  degree of danger in, 30–31
  as protected, 41
  worst-case scenario for, 30–31

Tamil Nadu, India, 356
Technical Report 049, of U.S. Fire
 Administration (USFA), 36
tenable smoke, 107, 127, 279
theaters
 as enclosed structures, 25
 in Wisconsin, 3
thermal imaging camera (TIC)
 for backdraft, 121
 in basements, 162, 241, 245, 246, 253
 in Carthage, Missouri, 306, 307
 in ceiling spaces, 222, 274
 in Charleston, South Carolina, 325, 335
 construction type and, 222
 in fire through roof, 346
 firefighter disorientation and, 352, 353
 for flashover, 129
 in high-rise structures, 169
 in large structures, 257, 258–259, 260, 268, 281, 282, 283
 in Phoenix, Arizona, 319
 for prolonged zero-visibility conditions, 174, 175
 ventilation and, 230
 in Worcester, Massachusetts, 297, 301
360-degree walk-around
 for basements, 54–55, 241, 245, 251
 Charleston, South Carolina and, 329
 flashover and, 129
 for large structures, 257, 258, 281, 289
 Memphis, Tennessee and, 303
 standard operating guidelines for, 307–308
 for wind-driven fires, 160, 163
 Worcester, Massachusetts and, 299–300
three-story structures
 as dangerous, 45
 as enclosed structures, 44
 as extremely dangerous, 33, 34
 fall height criteria for, 33, 34
 in Fall River, Massachusetts, 39–40
Tokyo, Japan, 356

Toledo, Ohio, Mayday incident in, 4
townhouses
 basements in, 53
 in Washington, DC, 62
truck company
 flashover and, 127
 in large structures, 284, 290
trusses. *See also* steel trusses; wooden trusses
 in attack spaces, 222
 bowstring, 225
 in Carthage, Missouri, 309
 in ceiling spaces, in attack spaces, 222
 firefighter disorientation and, 352
 in Phoenix, Arizona, 315
 scissor, 225
two-in/two-out rule, of Occupational Safety and Health Administration, 184
two-story garden apartment, flashover in, 20
two-story residential structure, 13–15, 14
 in residential districts, 18, 19
two-story retail structure
 in Annapolis, Maryland, 13
 as opened-structure, 15
two-story structures
 fall height criteria for, 33
 New York City Fire Department at, 134
 as opened structures, 15
 vacant, 75, 76
 walkout basement of, 50

U

Underwriters Laboratories (UL), 134–137
 on opened structures, 277
 on precooling, 232
 on self-venting, 140
 on wind-driven fires, 163
unknown–known guidelines (U–K)
 in Carthage, Missouri, 307
 in Charleston, South Carolina, 328

in Phoenix, Arizona, 315
standard operating guidelines and, 275–292
for Worcester, Massachusetts, 298–300
unprotected
apartment buildings as, 33, 44
basements as, 62, 236
at Carthage, Missouri, 305–308
at Charleston, South Carolina, 320, 328
defensive attack and, 258
degree of danger and, 33
dispatchers and, 319
enclosed structures as, 4, 5, 21, 23, 35–36, 176, 289, 338, 349–350, 355–357
extremely dangerous and, 21, 23, 28, 29
fall height criteria and, 33
flashover and, 131, 249
at Goldsboro, North Carolina, 68
high-rise structures as, 36, 44
I-joists and, 227
large structures as, 5, 29, 67, 68, 70, 75, 114, 289, 338
at Memphis, Tennessee, 301–302
opened/enclosed structures as, 64
at Phoenix, Arizona, 315
residential structures as, 12, 28
ventilation and, 230
wooden trusses and, 223
U.S. Fire Administration (USFA), 2
Analytical Database of, 108–111
Technical Report 049 of, 36
"U.S. Firefighter Fatality Retrospective Study" of, 97
"U.S. Firefighter Disorientation Study 1979–2001" (Mora), 6–8, 93, 108
"U.S. Firefighter Fatality Retrospective Study," of U.S. Fire Administration, 97
utility disconnect, 284, 290

V

vacant commercial structure
backdraft in, 123–124
with basement, 60–61
vacant duplex residence, rollover in, 170–171
vacant four-story building, as enclosed structure, 24
vacant structures
firefighter disorientation and, 94
wind-driven fire action plan at, 159
wind-driven fires in, 150–151
vacant two-story structure, as extremely dangerous, 75, 76
vacant warehouses, primary search and, 301
ventilation. *See also* vertical ventilation
backdraft and, 118, 122
into basements, 242, 246
burglar bars and, 24
for conversion steam, 137
coordination of, 231–232
doors for, 15
enclosed structures and, 21–25
fire dynamics and, 132
firefighter disorientation and, 229–232
flashover and, 133–134, 230
horizontal, 229
hose evolution and, 183
hydraulic, 230
in large structures, 28–29, 68
in opened structures, 19
in Phoenix, Arizona, 312
for subway stations and tunnels, 42
in tall structures, 30
unprotected and, 230
wind control devices and, 142
of wind-driven fires, 159–160, 230–231, 309
windows for, 15
in Worcester, Massachusetts, 300
for working fires, 231
vertical ventilation, 229
in Charleston, South Carolina, 343–344

381

construction type and, 231
in enclosed structures, 232, 343–344
fuel loading and, 343
in grocery stores, 232
wind-driven fire and, 343–344
in Worcester, Massachusetts, 297
Victoria, Australia, 357
vision difficulties. *See also* prolonged zero-visibility conditions; zero-visibility
in wind-driven fires, 153–154

W

walkout basement, of two-story structure, 50
walls, as wind control devices, 146–147
Walnus, Dennis, 154–155
warehouses
as enclosed structures, 40
in Salisbury, North Carolina, 62, 344
vacant, 301
Washington, DC, townhouses in, 62
water damage
balanced concern for, 129–130
from flashover contingency plan, 128, 130–131
from large structures, 292–293
size-up and, 106
Wayne Westland, Illinois, roof collapse in, 4
West Virginia, mobile home in, 153
West Warwick, Rhode Island, 356
Wharton, Texas, 5
"What Research Tells Us About the Modern Fireground" (Kerber), 137–138
Wholesale Fire & Rescue Ltd. (WFR), 255
wind control devices (WCDs), 140–142
ceilings as, 146–147
doors as, 146–147
firefighter disorientation and, 353
in high-rise structures, 147, 150–151, 168
walls as, 146–147

"Wind Driven Fire Research: Hazards and Tactics" (Madrzykowski and Kerber), 142–143
wind-driven fires, 139–170
action plan for, 159–163, 308
aggressive interior attack in, 152, 161
attic spaces and, 164–166
avoidance of, 142–143
in basements, 167, 240–241
breaching in, 147
breathing difficulty in, 153–154
in Carthage, Missouri, 305–309
ceiling spaces in, 164, 309
in Charleston, South Carolina, 329
in Chicago, 152
dispatchers and, 152, 159, 163, 308, 315–316
doors in, 144–146, 309
in enclosed structures, 27, 344
fire curtains for, 161
firefighter disorientation with, 89–90
floor collapse and, 344
flow path in, 133, 146, 150, 159, 164
in grocery stores, 157
hallways and, 103, 145, 148, 168
in high-rise structures, 143–152, 168–170
in Houston, Texas, 103, 152, 168
knock down in, 161
in large structures, 70
master stream in, 147
in mobile home, 152
in Muncie, Illinois, 3
in New York City, 152
in Oak Park, Michigan, 154–155, 158
in opened structures, 27
in Phoenix, Arizona, 157, 232, 310–320
in residential structures, 152
straight stream nozzles in, 149
ventilation of, 159–160, 230–231, 309
vertical ventilation and, 343–344

windows. *See also* boarded doors and/or windows; cinder block over windows; plywood sheeting, over windows
  backdraft and, 116, 124
  in basements, 53, 56, 242, 253
  in Charleston, South Carolina, 338–339
  enclosed structures and, 21–25
  for evacuation, 14, 16–17
  fire dynamics and, 132
  flow path and, 132
  in high-rise structures, 169
  in large structures, 268
  with metal frames, 17
  in opened structures, 12–20
  for ventilation, 15
Wisconsin, theater in, 3
Wolverhampton, UK, 349–350
wooden trusses
  dispatchers and, 308
  unprotected and, 223
Worcester, Massachusetts
  Cold Storage Warehouse in, 83, 102, 296–301
  handlines in, 297
  primary search in, 297
  prolonged zero-visibility conditions in, 297
  size-up and, 299–300
  360-degree walk-around and, 299–300
  vertical ventilation in, 297
  worst-case scenario in, 297
  zero-visibility in, 297
working fires
  doors in, 304
  firefighter disorientation in, 11
  with fuel loading, 27
  in New York City, 33
  in opened structures, 54
  ventilation for, 231
worst-case scenario
  for aggressive interior attacks, 27
  attack-supply evolution and, 186
  for enclosed structures, 27–28
  firefighter disorientation and, 352
  for opened structures, 26–27
  for tall structures, 30–31
  in Worcester, Massachusetts, 297

Z

zero-visibility, 79–80. *See also* prolonged zero-visibility conditions
  backdraft and, 123
  in basements, 64, 242–243
  in Carthage, Missouri, 306
  defined, 81
  in Fall River, Massachusetts, 39
  flashover and, 130
  handlines and, 96
  in high-rise stairwells, 66
  in large structures, 29, 69, 260, 265–266
  in opened structures, 27
  in Phoenix, Arizona, 313
  piercing nozzles and, 256
  in subway stations and tunnels, 42
  in Worcester, Massachusetts, 297